Precious Metals:
Minerals and Health

Carol Jong, Ph.D., R.D.

A Biomed Publication
5801 Christie Avenue, Suite 400
Emeryville, California 94608 USA
Tel: (510) 450-1650
Fax: (510) 652-1859

To obtain more information about Biomed's products and services, please feel free to contact the company. The address is: Biomed, 5801 Christie Avenue, Suite 400, Emeryville, California 94608 USA. Tel: (510) 450-1650; Fax: (510) 652-1859.

TABLE OF CONTENTS

LISTING OF TABLES

LISTING OF HOT TIPS

ABOUT THE AUTHOR

CAROL JONG, Ph.D., R.D.

Carol Jong, Ph.D., R.D. obtained her doctoral degree in nutrition from the University of California at Davis. She received her bachelor of science in public health from the University of California at Los Angeles. She has extensive academic, research, and professional experience in nutritional metabolism and "nutraceuticals" (nutrients which prevent and treat disease).

In the fields of nutrition and pharmacology, Dr. Jong has presented seminars throughout the United States. She has also lectured extensively abroad. Her seminars have trained and educated nurses, physicians, dentists, pharmacists, dietitians, and other health professionals. She teaches at the university level and has been a featured nutrition expert in the news media.

WORDS FROM THE AUTHOR

As I travel throughout various countries, providing nutrition and health seminars, I realize that there is much interest and concern about minerals as essential nutrients. I also realize that misconceptions and inaccuracies about mineral supplements and products abound. Exaggerated claims and extrapolated science have steered sound nutrition and common sense astray. In this book, I will examine and explore this contemporary and fascinating area of nutrition and develop and enrich your knowledge of nutrition, metabolism, and wellness. Our exploration will be based on contemporary knowledge and empirical research. I share this information with you, the reader, in hopes of educating you to live smarter, better, and longer.

--Carol Jong, Ph.D., R.D.

CHAPTER ONE

Can We Talk? The Basics

Minerals: Past and Present

Past: The Elusive Ones

Did you know that to maintain good health the body requires at least 45 dietary nutrients and that nearly half of these essential nutrients are minerals? The large number of minerals and their diverse roles in metabolism have made them, by comparison to other nutrients (carbohydrates, fat, protein, and vitamins), elusive chemicals which have received less than their fair share of attention. Fortunately, research is bringing minerals onto the horizon and recognizing these unique nutrients and their specific roles in disease prevention and overall health. Traverse these pages and arm yourself with the information you need to make smart choices towards health.

Present: Cowboys on the "Nutraceutical" Frontier

An important evolution in nutrition has brought minerals onto the "nutraceutical" (nutrients which prevent and/or treat disease) frontier, undoubtedly a prelude to a new understanding. In the next few sections, you will understand how minerals steer and direct metabolism, thereby, defining your well-being. You're invited not to simply witness this evolution, but, by embracing the information, live it! Knowledge is power! As you increase your knowledge and awareness of the role minerals play in determining good health, you gain the confidence needed to make the right decisions for your body. You also got a future of good health.

Minerals: What Are They, Anyway?

Textbook definition: inorganic nutrients that act as cofactors to influence "metabolism." What is metabolism? Envision the body as a test tube of chemical reactions which cause a nutrient to be transformed into another chemical. The final product affects our metabolism and defines our body's metabolic outcome, our physiological state, and our health. Minerals have a profound and ubiquitous effect on metabolism because they influence enzymes, and therefore, chemical reactions and therefore, the body's metabolism. Minerals have been implicated in influencing a diverse array of systems, from energy balance to immune function to free radical formation to gene expression. Minerals are magicians with the power to cause nutrient metamorphosis, with each mineral having a unique and very specific effect upon the body.

Who's Driving the Metabolism Bus?

Metabolic reactions are under regulation; that is, they are not a random plethora of chemical reactions running wild, but rather a defined and conducted orchestration. These reactions are under the control of enzymes. Enzymes enable reactions to occur quickly and efficiently. Furthermore, enzymes, often require a mineral cofactor to function. The inactive form of an enzyme is called an apoenzyme. When a mineral cofactor and apoenzyme combine, the complex, now active, catalyzes metabolic reactions.

Hot Tip:
Did you know that...
Zinc is a cofactor for over 300 enzymes?

CHAPTER TWO

How Much and Who Says?

Minerals: Big, Bigger, Biggest

Minerals are classified as either major or trace, depending on the amount needed by the body each day. Generally speaking, the need for major minerals is 100 mg (approximately 1/50 teaspoon) or more per day. The need for trace minerals is lower. For all their metabolic importance, we need only approximately ¼ teaspoon of minerals each day for adequate nutrition! Major and trace minerals are listed in Table I.

What Is Essential?

The essentiality of some minerals remains elusive. Table I describes the minerals which have been proven to be either essential in the human diet (see columns labeled "Essential Major Minerals" and "Essential Trace Minerals") or have questionable essentiality (see column labeled "Possibly Essential").

Table I. Essential and Non-Essential Minerals.

Essential Major Minerals	Essential Trace Minerals	Possibly Essential Minerals
Calcium	Chromium	Arsenic
Chloride	Cobalt	Boron
Magnesium	Copper	Cadmium
Phosphorus	Fluoride	Lithium
Potassium	Iodine	
Sodium	Iron	
Sulfur	Manganese	
	Molybdenum	
	Nickel	
	Selenium	
	Silicon	
	Tin	
	Vanadium	
	Zinc	

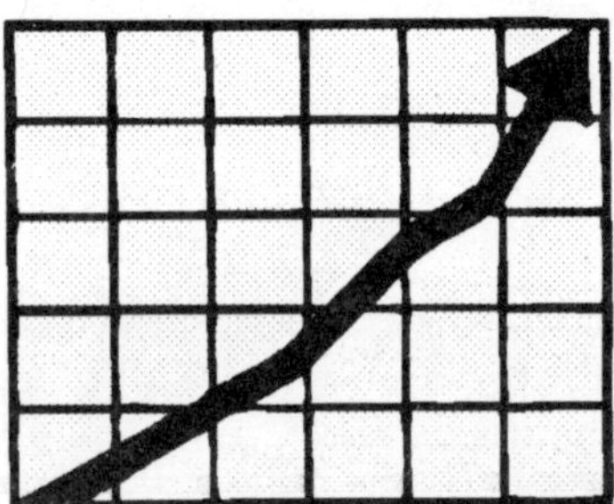

How Much and Who Says?

For most essential nutrients, there is a Recommended Dietary Intake (RDI). The best known of the RDIs is the Recommended Dietary Allowance (RDA). The RDAs are recommendations developed by the Food and Nutrition Board. The board is made up of nutrition experts appointed by the United States Department of Agriculture. RDAs are directed toward healthy individuals and are intended to prevent nutrient deficiency diseases and to maintain overall health by providing adequate nutrition for basic body functions, such as growth, maintenance, and repair of tissues.

Table II. Recommended Dietary Allowance for Selected Minerals.

Mineral	RDA Male (25 to 50 years)	RDA Female (25 to 50 years)	RDA Male (51+ years)	RDA Female (51+ years)
Calcium (mg)*	800	800	800	800
Phosphorus (mg)	800	800	800	800
Magnesium (mg)	350	280	350	280
Iron (mg)	10	15	10	10
Zinc (mg)	15	20	15	12
Iodide (ug) **	150	150	150	150

***mg: milligrams, which is 1/1000 of a gram.**
**** ug: micrograms, which is 1/1000 of a milligram.**

Sometimes available scientific research is limited and/or inconclusive, and RDAs cannot be established. Consequently,

RDIs for some nutrients are described by the Estimated Safe and Adequate Daily Dietary Intake (ESADDI) and Estimated Minimum Requirement (EMR) which are listed in Tables III and IV, respectively.

Table III. Estimated Safe & Adequate Daily Dietary Intake for Minerals.

Mineral	ESADDI for Adults
Copper (mg)	1.5 to 3
Manganese (mg)	2 to 5
Fluoride (mg)	1.5 to 4
Chromium (ug)	50 to 200
Molybdenum (ug)	75 to 250

Table IV. Estimated Minimum Requirement for Minerals.

Mineral	EMR for Adults (> 18 years)
Sodium (mg)	500
Phosphorus (mg)	500
Potassium (mg)	2000

As its name suggests, the ESADDI is a judgment of the safe and adequate intake for a given nutrient. The EMR is the level estimated to prevent nutritional deficiency.

CHAPTER THREE

Getting Your Minerals ... And Keeping Them

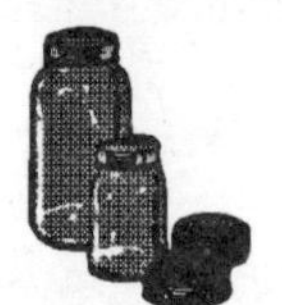

If a Little Is Good, Then More Must Be...Better.

Remember our discussion about the RDAs? Do you know that some health professionals and researchers believe that the RDAs for some nutrients provide adequate nutriture, but not necessarily optimal nutriture. What's the difference?

Adequate

The RDAs are recommendations to prevent classical nutrient deficiency diseases and to maintain growth, maintenance, and repair of tissues. They do not consider the relationship between a nutrient and prevention of chronic diseases such as coronary heart disease and cancer. The RDAs are nutrient requirements for generally healthy persons. The RDAs may not be adequate for the following conditions or individuals:

- Illness
- Chronic use of medications
- Elderly
- Athletes, or
- The presence of other factors that can alter nutrient requirements.

RDAs are recommendations for maintenance of nutrient status and may be inadequate if nutrients stores must be repleted because of illness, poor dietary habits, and/or increased nutrient requirements.

Optimal

Optimal intake is the level of a nutrient which will provide maximal health, the lowest risk of disease (in the context of genetic influences), and optimal performance for athletes. Generally speaking, the optimal level exceeds the adequate level. For instance, the calcium RDA for adults is 800 mg. As a nutraceutical, a daily calcium intake of 1200 to 1500 mg reduces osteoporosis risk; 800 to 1200 mg decreases hypertension; and 1200 to 1500 mg lowers low-density lipoprotien (LDL) cholesterol levels.

Adequate or Optimal? Which Will It Be?

Usually the quantity necessary to achieve optimal nutriture is greater than the quantity necessary to achieve adequate nutriture.

If a Little Is Good, Then More Must Be...Worse.

Minerals and their specific effects are often considered individually, but that is certainly not how the body "sees" nutrients. Remember that nutrients are obtained from food (hopefully), or even supplements which contain multiple vitamins and minerals. The complex interactions which occur among these nutrients dictates their bioavailability to the gut and to whole body metabolism. So, the plot thickens...if a little is good, then more must be better, but not always. Read on to understand this caveat.

Bioavailability: A Gut Feeling.

Nutrient-nutrient interaction. How does one nutrient in the diet affect the metabolism of another nutrient? Out of all the chemicals in the diet, the bioavailability of minerals is most influenced by the presence of other minerals and nutrients. Since many minerals compete with one another for absorption in the gut, an overabundance of one mineral can produce a deficiency of another. When one nutrient or condition creates a deficiency of another nutrient, this is known as a secondary deficiency.

Secondary deficiencies of iron, zinc, and copper are likely if one of these minerals is excessive in the diet. Especially critical to determining whether an interaction has an inhibitory or enhancing effect upon bioavailability is the <u>proportion</u> of minerals which are present. Consider Table V to see which minerals have potential interactions and to see the maximum proportion (see column labeled "Maximum Proportion to Avoid Interaction") wherein no detrimental interaction occurs.

Are You Interacting? Using Table V.

To determine whether there is potential interaction between and among minerals, let's examine the minerals which have potential interactions, then determine their proportion in your diet. For this table, we will, for two reasons, focus upon mineral intake from supplements:

- Most minerals interact if in excess, which is more likely to occur through supplement usage rather than from regular consumption of foods.
- Information about mineral intake is most readily available from supplement labels. If you are taking mineral supplements, refer to the nutrition label to determine the quantity of a given mineral. If you know your mineral intake from your diet and food (perhaps you have a nutrient analysis of your diet, which details mineral intakes from food), then this extra information would make the exercise in Table V even more informative.

Refer to the mineral in Column A of the table; write down your intake of this mineral in Column D.

Refer to the mineral in Column B of the table; write down your intake of this mineral in Column E.

Remember that all quantities should be written down as "milligrams" (mg). If a quantity is listed on the nutrition label in "micrograms" (ug), simply convert ug into mg in the following manner:

______ ug divided by 1000 = ______ mg

Example:

2500 ug divided by 1000 = **2.5** mg

Divide the quantity in Column D by the quantity in Column E. Write this number in Column F.

Does the number in Column F exceed the number in Column C? If yes, then check the appropriate box in Column G. If "no," then check the appropriate box in Column G.

A "yes" answer means that an excessive quantity of the

mineral in Column A may be interfering with the bioavailability of the mineral in Column B. Recommendation: Cut back on your intake of Column A mineral to achieve a more optimal proportion and reduce the potential for inadequate nutriture and a secondary deficiency of the interacting mineral.

Hot Tip:
What's an Ion? What's an Electrolyte?

Minerals are ions when they carry a positive or negative charge. Some examples of positively and negatively charged minerals/ions are:

Mineral/Ion	Charge
Calcium	2+
Chloride	1-
Chromium	3+
Copper	2+
Iron	2+ or 3+
Magnesium	2+
Manganese	2+
Zinc	2+

Can you see why an excess of one mineral can compromise the bioavailability of another? An excess of one mineral can "outcompete" another mineral for absorption, since there is a limited and finite number of transport proteins.

When minerals do not have a positive or negative charge, they are "neutral" (i.e., have no charge). In the neutral state, minerals are called electrolytes. In nature, however, minerals generally do carry a charge (refer to the chart above). When in the charged state, minerals are called ions.

Note that many minerals have a charge of "2+." Hence, these minerals are similar in their chemistry. In fact, these minerals are so similar that they actually compete with one another in the intestine. When we eat food, we also consume minerals. These minerals eventually enter the intestine and "wait" for absorption into the bloodstream. If minerals are similar, they "look" the same to the intestine and can compete for absorption into the bloodstream, possibly resulting in an "imbalance" of nutrients.

Table V. Mineral Interactions.

Column A	Column B	Column C	Column D	Column E	Column F	Column G	
Origin of Interaction (X)	Mineral Interaction (Y)	Maximum Proportion to Avoid Interaction (X/Y, Mg/Mg)	Your Intake of Mineral X (Mg)	Your Intake of Mineral Y(Mg)	Interaction Ratio (X Divided by Y)	Interaction Potential?	
						Yes	No
Excess Calcium	Zinc	< 200					
Excess Iron	Zinc	< 20					
	Manganese	< 5					
	Fluoride	< 87					
Excess Zinc	Iron	< 2					
	Copper	< 15					
	Manganese	< 8					
	Fluoride	< 75					
Excess Copper	Zinc	< 1					
	Selenium	< 3300					
Excess Manganese	Zinc	< 8					
	Selenium	< 3300					
Excess Fluoride	Calcium	< 1.02					
	Phosphorus	< 0.06					
	Magnesium	< 0.2					
Excess Iodine	Fluoride	< 0.75					

Hot Topic: Calcium vs. Iron: What's a Woman to Do?

To prevent osteoporosis, women are consuming calcium supplements and high-calcium foods; *but listen to the rest of the story.* High intake of dietary calcium can inhibit iron absorption, particularly if both are present in the same meal. Optimal nutriture of both minerals is critical because of the nutraceutical effects of calcium (i.e., the reduction of the risk of osteoporosis, hypertension, coronary heart disease, and colon cancer) and biological effects of iron (i.e., oxygen transport, energy production, and mental performance). So how does one insure optimal nutriture of both minerals?

Researchers have shown that when high-calcium foods, such as milk and cheese, are present in a meal, iron absorption is much lower than when dietary calcium is not present.

Experiments on rats showed that high dietary levels of calcium suppressed iron absorption. Rats fed high levels of calcium carbonate (the most common type of calcium supplement) showed a sharp drop in iron absorption. Interestingly, this situation occurred in iron-deficient rats that were in a state in which iron absorption should have been very high.

What can we learn from this? Calcium and iron supplements should not be consumed simultaneously. Fortunately, calcium supplements are typically available as a single nutrient supplement, rather than combined with other minerals. Think about how large calcium supplement pills are. Two large tablets typically contain the USRDA for calcium (1200 mg). The USRDA is a special RDA set by the Federal Government. If calcium were to be integrated into a multi-vitamin/mineral supplement, think about what a "horsepill" this would be!

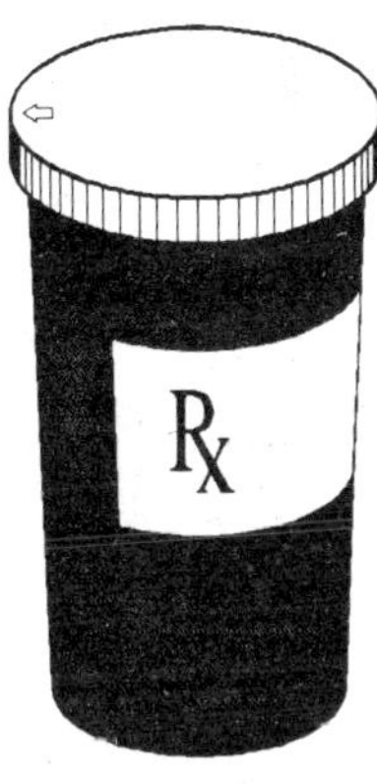

Also remember that most nutritional supplements should be consumed with meals. (Iron supplements are an exception to this rule. These should be consumed on an empty stomach, since many food components can interfere with iron's bioavailability.) Why? When we eat food, the body undergoes a "cephalic phase," that is, a phase in which the anticipation and consumption of food stimulate bodily changes such as salivation (Remember Pavlov's dogs?), stomach-acid production, and digestive-enzyme secretion. All of these enhance the bioavailability of nutrients.

Ideally, calcium and iron should be obtained from food sources, rather than solely from supplements. Tables VI and VII contain a listing of calcium-rich and iron-rich foods, respectively. Note that the foods in these tables are high in their nutrient densities for calcium and iron; that is, these foods provide the greatest calcium or iron for a given quantity of calories.

Hot Tip:
What to Believe?

One liquid mineral supplement recommends the following: "Take between meals, away from food or other supplements." Actually, with the exception of iron, minerals and essential nutrients are best absorbed in the presence of food.

Table VI. Calcium-Rich Foods.

Calcium-Rich Food	Calcium Content (mg/serving)	Serving Size	Calories per Serving
Cheese manicotti	296	1 oz	29
Cheese stuffed shells	248	1 oz	28
Cheese ravioli	336	1 oz	54
Spinach, cooked	122	0.5 cup	21
Collards, frozen, chopped, cooked	136	0.5 cup	25
Cheddar cheese, non-fat	243	1 oz	68
Milk, 2% fat	297	1 cup	86
Chocolate milk, 2% fat	284	1 cup	158
Cottage cheese, 2% fat	161	1 cup	211

Hot Tip:
Calcium & Food Labels

The Food and Drug Administration (FDA) describes the calcium content of food according to the following definitions:

- "High-calcium": 200 mg or more per serving.
- "Good source of calcium": 100 to 200 mg per serving.
- "More or added calcium": 100 mg more per serving than the standard recipe

Table VII. Iron-Rich Foods.

Iron-Rich Food	Iron Content (mg/serving)	Serving Size	Calories per Serving
Cereal, 40% bran flakes	25	1 cup	127
Clams, cooked, moist heat	24	3 oz	126
Oatmeal, dry	28	1 cup	153
Cereals, fortified	21	1 cup	120
Spinach, cooked	4.6	1 cup	21
Cereal, raisin bran	25	1 cup	165
Oysters, raw	4.9	3 oz	50
Asparagus, canned	4.4	1 cup	46

...So now you know the rest of the story.

Antinutrients.

Antinutrients are dietary components which impair the absorption and metabolism of other nutrients. Thus far, we have already looked at antinutrients. Minerals causing a nutrient inadequacy or secondary deficiency can be labeled as antinutrients. Read on to see which nutrients have antinutrient effects upon mineral metabolism.

Fat and Calcium Soaps.

Did you know that calcium binds with fat in the intestine? Did you know that when these complexes are formed, neither calcium nor fat can be absorbed? These complexes appear sudsy and hence, are called "calcium soaps." Calcium soaps have various consequences on the body:

- ❑ ...a detrimental effect
  Populations which have high fat intakes also exhibit greater likelihood of osteoporosis. (Because of decreased bioavailability of calcium.)

- ❑ ...or a beneficial effect
  A calcium intake of 1200 to 2200 mg per day significantly reduces LDL cholesterol. This cholesterol is low-density lipoprotein cholesterol or "bad" cholesterol. LDL cholesterol levels drop because of decreased absorption of dietary fat.

- ❖ A calcium intake of 1500 to 2000 mg per day significantly reduces colon cancer risk by forming calcium complexes with fatty acids so that they cannot irritate, damage, or cause cell abnormalities in the colon.

Phytic Acid.

Phytic acid is a constituent of plant fibers and is typically found in cereals and grains. Phytic acid is greatly attracted to minerals, particularly calcium, magnesium, iron, copper, and zinc. When phytic acid complexes form with these nutrients, minerals are not bioavailable.

But wait. Haven't we been told to increase our consumption of cereals and whole grains, foods rich in phytic acid? Does heeding this recommendation put us at risk for inadequacies or secondary deficiencies of minerals? No. In American culture, grains are processed in a manner that reduces the phytic acid content in grain products. During the production of grain products, yeast is used as a leavening agent. During the leavening process, enzymes produced by the yeast cells degrade and destroy the phytic acid. Hence, Americans are less susceptible to the antinutrient effects of phytic acid. Diets that consist of large quantities of unleavened grain products (i.e., Middle Eastern) may experience a greater risk of a secondary inadequacy or deficiency of some minerals. Some examples of unleavened grain products include: matzo and nan, bread staples which are native to the Jewish and Indian cultures, respectively.

Fiber.

Dietary fiber can be an antinutrient because of its ability to form non-absorbable complexes with minerals. There are two types of dietary fiber: soluble and insoluble fiber, so named because of soluble fiber's ability and insoluble fiber's inability to dissolve in water. Insoluble fiber, in grains such as wheat and rice, form complexes with minerals. (Dietary fiber is found only in plant foods; diets which are high in plant foods contain a large quantity of dietary fiber.) Hence, there is some concern about high-fiber diets, such as vegan (plant products only) diets. Whether vegans receive bioavailable minerals and adequate nutriture has been questioned. See the section entitled "No Meat For Me, Thank You...The Vegan" for a discussion of this issue.

Coffee, Tea, and Caffeine.

Does coffee drinking cause calcium loss from the body? Most studies (but not all) have failed to show a dose-response relationship between coffee consumption and increased risk of osteoporosis. In other words, most studies (but not all) have failed to show that the more coffee one drinks, the greater the calcium loss from the body.

- ❑ First, there seems to be a threshold level of caffeine intake at which calcium bone loss increases. In a group of postmenopausal women, those with the highest caffeine intake (greater than 450 mg per day, or approximately 3-1/2 or more cups of brewed, regular-strength coffee) had significantly more bone loss than women consuming less caffeine. To determine whether you are in the high-risk group for caffeine intake, refer to Table VIII to see the caffeine content of various foods.

Table VIII. Caffeine-Rich Foods.

Caffeine-Rich Food	Caffeine Content (mg/serving)	Serving Size	Calories per Serving
Coffee, brewed	103	6 oz	4
Coffee, instant, regular	57	6 oz	4
Tea, brewed	47	8 oz	2
Tea, instant	31	8 oz	2
Dr. Pepper	61	12 oz	151
Coca-Cola	65	12 oz	152
Pepsi	36	12 oz	152
Mountain Dew	55	12 oz	165
Cocoa *	13	6 oz	250

* Cocoa and therefore chocolate contains a minimal amount of caffeine. The primary stimulant in cocoa is "theobromine," a compound related to caffeine. There is an average of 250 mg of theobromine in 6 ounces of cocoa, which explains the stimulating effects of chocolate.

- ❑ Secondly, lower caffeine intake (equivalent to 2 to 3 cups of brewed coffee) along with low calcium intake (less than the

RDA of 800 mg) may accelerate bone loss. Use Table VI along with the Rule of 300 described below to determine whether you fall into this risk category.

Hot Tip:
The Rule of 300.

There is a quick and easy method to "guestimate" your daily calcium intake. This method is called "The Rule of 300." (The number 300 is used because convenient serving sizes of calcium-rich foods generally contain about 300 mg of calcium). The Rule of 300 requires the use of a special chart and doing some simple arithmetic. Table IXA. below provides an example of the chart.

Table IXA. The Rule of 300.

Balanced diet?	Yes	X 300 mg calcium	=
Balanced diet?	No	X 0 mg calcium	=
Number of high-calcium food servings*		X 300 mg calcium	=
Guesstimated calcium intake (mg/day)			

Hot Topic
By the Way:
What's a Serving?

Listed below are high-calcium foods and convenient serving sizes. Each serving size contains approximately 300 mg of calcium. Also below is Figure I, the Food Pyramid Guide of the United States Department of Agriculture.

High Calcium Food	One Serving Size
Yogurt	1 cup
Milk	1 cup
Tofu	2 squares
Ice cream	2 cups
Cottage cheese	2 cups
Cheese	3 oz or 3 slices

Figure I. USDA Food Guide Pyramid.

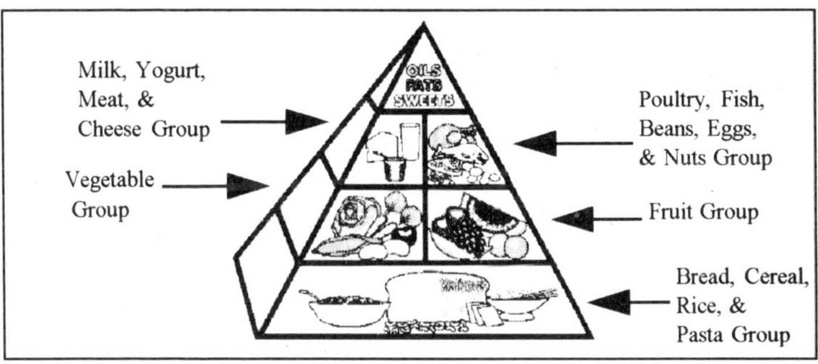

Example #1: The Rule of 300.

Let's use this information in an example.

Let's say you have a fairly "balanced" diet (i.e., your diet is similar to the USDA Food Pyramid Guide). As far as dairy products are concerned, you routinely consume ½ cup of yogurt and 1 cup of milk each day.

Table IXB. Example #1 of The Rule of 300.

Balanced diet?	Yes X	X 300 mg calcium	= 300 mg
Balanced diet?	No	X 0 mg calcium	=
Number of high-calcium food servings	1.5	X 300 mg calcium	= 450 mg
Guesstimated calcium intake (mg/day)			750 mg

Example #2: The Rule of 300.
Let's use this information in an example. Let's say you do not have a fairly "balanced" diet (i.e., your diet doesn't look anything like the USDA Food Guide Pyramid). As far as dairy products are concerned, you routinely consume ½ cup of yogurt and 1 cup of milk each day.

Table IXC. Example #2 of The Rule of 300.

Balanced Diet?	Yes	X 300 mg calcium	=
Balanced Diet?	No X	X 0 mg calcium	= 0 mg
Number of high-calcium food servings	1.5	X 300 mg calcium	= 450 mg
Guesstimated calcium intake (mg/day)			450 mg

Now use The Rule of 300 for yourself. You can use Table IXA. on Page 16. If after completing this calculation, you determine that your calcium intake is at least 800 mg, then congratulations! You satisfy the RDA for calcium.

If, after completing this calculation, you determine that your calcium intake is at least 1200 mg, then--better yet--you satisfy the USRDA, a special RDA set by the Federal Government. (The USRDA is the RDA for adults, male or female, whichever is greater.) For calcium, the USRDA of 1200 mg is, in terms of preventing osteoporosis, better than the RDA of 800 mg.

Alcohol.

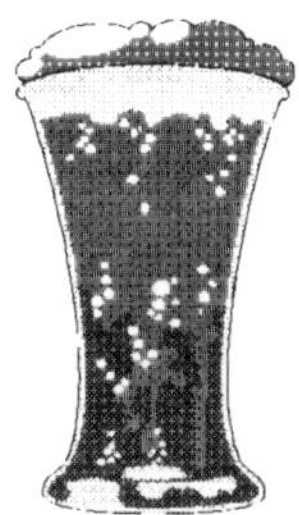

In America, alcohol is the leading cause of secondary nutrient deficiencies. Alcohol reduces food intake, and therefore nutrient intake, and alcohol damages the lining of the intestine, thereby interfering with the absorption of essential nutrients. Excessive alcohol intake also causes greater urinary loss of magnesium, potassium, and zinc.

Given these effects of alcohol upon nutrient metabolism, one would expect that alcohol consumption compromises calcium nutriture, increasing the risk of osteoporosis. Is alcohol consumption a risk factor for osteoporosis? Several studies have attempted to resolve this question. To date, there is not a clear dose-response relationship between alcohol and bone mineral density. Surprisingly, research has shown that individuals who drink at least 7 ounces of alcohol per week (approximately 4 ounces of wine, 1 beer, or 1 ounce of 80-proof alcohol per day) actually have higher bone densities than light drinkers (less than 1 ounce of alcohol per day). Interestingly, this effect was more pronounced in women than men. Researchers believe that alcohol increases bond density by augmenting the effect of estrogen, which enhances bone mineral deposition.

Hot Tip:
Alcohol: What's Enough & How Much Is Too Much?

You've probably heard that "moderate" alcohol consumption may reduce the risk of coronary heart disease, but what is moderate? One ounce of alcohol (the amount in 1 glass of wine, 1 beer, or 1 ounce of 80-proof alcohol) per day is approximated as "moderate." Of course, this recommendation is not intended to encourage individuals to begin drinking alcohol if they are currently not doing so, since alcohol can be addictive and a source of excess calories. Rather, this recommendation is intended to provide guidelines for the individual who already consumes alcohol.

Sodium vs. Calcium: Who's Winning the War?

In addition to raising blood pressure, dietary sodium has been implicated in a number of other diseases, including osteoporosis. Consider that salt intake has been associated with an increase in urinary calcium excretion (creating a predisposition to osteoporosis and kidney stones) and an increase in urinary hydroxyproline (the protein which strengthens bones). Given that for every 1100 mg of sodium there is a 203 mg urinary calcium loss. A typical sodium intake of 3000 to 5000 mg each day, can significantly reduce bone mineral density. For individuals consuming excess sodium, a net loss of calcium can significantly reduce bone density and increase osteoporosis risk. Interestingly, the converse is also true. Additional calcium can increase the excretion of sodium through the urine, a phenomenon which may help reduce blood pressure in sodium-sensitive individuals. For instance, for every 150 mg of calcium consumed, 20 mg of sodium is excreted in the urine. Table X provides a listing of sodium-rich foods which should be consumed in moderation. As a standard of reference, the Estimated Minimum Requirement for sodium is 500 mg per day.

Table X. Sodium-Rich Foods.

Sodium-Rich Food	Sodium Content (mg/serving)	Serving Size	Calories per Serving
Soup, French onion	2073	1 cup	100
Soup, chicken broth or bouillon	1484	1 cup	22
Soup, gazpacho	1183	1 cup	56
Soup, tomato vegetable	1146	1 cup	12
Soy sauce	1029	1 tbsp	10
Peppers, hot chili, red/green, canned	856	1 each	18
Tofu, salted and fermented	815	1 oz	33
Sauerkraut	780	0.5 cup	22
Pickle, cucumber, dill	771	1 slice	1

CHAPTER FOUR

Chelators, Colloids, & (a Little) Chemistry

Partners In Crime: Minerals and Their Chelators.

Have you ever seen a supplement label claiming that it contains only "chelated minerals?" Have you ever wondered what this means? Have you ever wondered if it is superior to other supplements? Have you ever wondered if it is worth the extra money? Well, wonder no more.

Chemically speaking, minerals do not exist alone. In nature, minerals are chemically bound to another compound. The other compound is called a "chelator" (The bound entity is called a "chelate"). A very common chelator used in calcium supplements is "carbonate." Hence, the supplement contains calcium-carbonate, which is a mineral-chelator complex. Chelators can differ. Calcium-citrate is another popular mineral-chelator complex in supplements. Since all minerals can exist only as mineral-chelator complexes, all supplements contain "chelated minerals."

...But not all chelators are created equally. A chelator can affect the bioavailability of a mineral. How? All minerals exist as part of a chelator complex, but in order for a mineral to be absorbed, it must first be removed from the chelator. The chemical bond which attaches the mineral to the chelator must be broken so that the free mineral can be absorbed. If the chemical bond is easily broken, then the mineral is readily absorbed.* Each mineral and chelator have different chemical bonds, and the bonds are broken which varying ease. Also, some chelators may change the environment of the digestive system so the bioavailability of the mineral is reduced or enhanced. For example, citrate is not only a chelator but also an acid which can increase the bioavailability of calcium.

Some minerals are marketed as being better absorbed

*An exception to this rule is chromium picolinate. This mineral-chelator complex is absorbed intact.

because they are chelated with amino acids, nucleic acids, or metabolic acids. Consider the following statement from the label of a trace mineral supplement:

"Provides an advanced delivery system for trace minerals. Microscopic particles of the minerals are organically bound to physiologically active amino acids or metabolic acids then uniformly suspended in a liquid colloidal system balanced to optimum pH."

Let's look at some of the characteristics of this supplement, as described in this statement:

- *"...organically bound..."*
  Inorganic vs. organic minerals. Is there a difference? Yes, chemically speaking. "Inorganic minerals" (i.e., does not contain the element carbon) are minerals in their elemental form, which are not bound to a chelator. "Organic minerals" are minerals which are bound to a chelator, which itself is organic (i.e., contains the element carbon). Thus, when minerals are bound to chelators such as amino acids and metabolic acids, as described in this statement, they are "organic."

- *"...physiologically active amino acids or metabolic acids..."*
  Our bodies use amino acids for various functions, and also produce some amino acids and metabolic acids. Because these are "physiologically active" (i.e., they have a role in metabolism), they do not necessarily enhance the bioavailability of minerals to which they are attached. As stated above, prior to absorption, minerals must be removed from their chelators (even if they are amino acids or metabolic acids) so that minerals will be in their elemental form necessary for absorption.

- *"...uniformly suspended in a liquid colloidal system..."*
  ...or stirred, not shaken.
  The mineral-chelator complexes are then mixed into a liquid, sometimes a sugar/water mixture.

- *"...balanced to optimal pH."*
  What is optimum pH? The pH of the digestive tract can vary from pH of 1 (very acidic!) to a pH of 7 (neutral). What

is optimal pH? Most minerals are best absorbed in an acid environment, which the stomach creates by producing a very strong acid, hydrochloric acid.

Hot Tip:
Chelation Therapy: Not the Same Thing

"Chelation therapy" has become more popular with the increasing interest and use of alternative medical therapies. In order to understand chelation therapy, we can use many of the basic principles discussed in the section above. Imagine that an individual has excess accumulation of toxic minerals, such as aluminum, arsenic, copper, lead, iron, etc. Chelators are administered to bind with these minerals and remove them from the body via the urine.

Suspended animation: Mineral Colloids Fact vs. Fiction.

The latest development in mineral supplementation is minerals suspended in liquid. These are called "colloidal minerals" and have the supposed benefit of supplying essential major and trace minerals with a uniquely high bioavailability. Let's examine these issues.

Manufacturers have been aggressive in informing the consumer about the need for essential minerals and supplements. It is true that the American diet is often inadequate in calcium, iron, zinc, copper, and chromium. Resultantly, mineral supplements may not only provide adequate, but optimal nutriture. However, more is not always better. Recall our previous discussion of secondary inadequacies and deficiencies induced by an excess of specific minerals.

Let's now examine mineral colloids, their chemical make-up and whether this affects their bioavailability. Colloid minerals are a mixture of water and mineral-containing clay. When water, the dispersing agent is mixed with clay, the minerals of the clay* become dispersed in the water and a colloidal mineral is created. When the minerals are dispersed in the mixture, they are still solid in form; they are suspended (but not dissolved) in liquid, and therefore, are not truly liquid in form.

* Why clay? Soil and other geological materials are an inexpensive source of dietary minerals. Elements found in the Earth's crust are listed in Table XI.

Table XI. Elements of The Earth's Crust.

Element	Content in Earth's Crust (% of mass)
Oxygen	46.5%
Silicon	28.0%
Aluminum	8.1%
Iron	5.1%
Calcium	3.5%
Sodium	3.0%
Potassium	2.5%
Magnesium	2.2%
Titanium	0.5%
Hydrogen	0.2%

Hot Topic: Are Our Soils Depleted of Minerals?

Many of our minerals come from plants (i.e. grains, cereals, fruits, and vegetables). As a plant grows, it concentrates and extracts soil-based minerals that it needs to survive, thrive, and produce food for us. Many think that because of overuse, U.S. soils are so depleted of minerals, that plants grown in these soils will be unable to deliver minerals needed for human nutrition.

Is this true? Typically, plants have mineral requirements for themselves. If these minerals are not present in the soil, then the plants could not survive, and we would not be eating them as food. We only eat plants which are healthy enough to survive and bear fruit, flowers, and other edible parts. These plants, by containing adequate minerals for their own survival, can contribute to our survival. Moreover, with our borderless agriculture, we import produce from many parts of the world. This gives us access and exposure to plants from many different soils. These imported plants, depending on the environmental and geological make-up of the area of origin, can result in the delivery of a wider variety of minerals, helping to insure that humans get their fair share of minerals

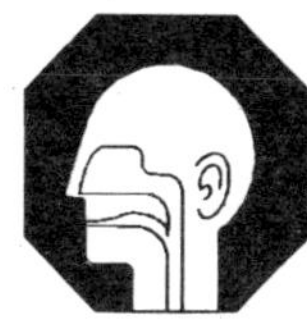

Body Composition...

The mineral composition of mineral colloidals varies among manufacturers. Remember that mineral colloidals are usually a suspension of clay or organic material of soil, which contains not only minerals, but also "elements," many of which are found in the body. In chemistry, minerals and elements are organized according to increasing size and weight. Some colloidal manufacturers claim that their supplements contain the first 92 elements (i.e., from the smallest element (1st) to the 92nd element)...but do they? And in what amount?

...Too Little.

In defining some of the terminology used to describe the mineral and element content of these supplements, we will see that the quantities of some minerals are insignificant and insufficient to supply optimal, let alone adequate nutrition.

- Parts per million (ppm).
  Consider breaking a metal paper clip into 1000 pieces. Then break one of those pieces into 1000 pieces. One of these pieces is equivalent to 1 ppm, relative to the whole paper clip.

- Parts per billion (ppb).
  Consider breaking a metal paper clip into 1000 pieces. Then break one of those pieces into 100,000 pieces. One of these pieces is equivalent to 1 ppb, relative to the whole paper clip.

- Not detectable (ND).
  When a supplement is analyzed, the content of that mineral or element may be too low to be measured by the equipment being used for the analysis. One can also presume that its mineral content is too low to nourish all the cells of the body.

So what is the mineral content of colloidal mineral supplements? Table XII describes the elemental content of three colloidal mineral products analyzed by an independent laboratory. All values in Table XII are expressed in ppm. Note the significant number of elements which are too insignificant to be detected (i.e., "ND").

Table XII. Element Composition of Colloidal Mineral Products.

Element	Product X	Product Y	Product Z
Major minerals			
Sodium	22,995	55.3	218.8
Potassium	97.4	ND	26.4
Calcium	ND	488	517.6
Magnesium	ND	402	694.6
Trace elements			
Boron	18.8	ND	8.5
Cobalt	ND	ND	2.1
Copper	1.3	2.63	0.554
Iodine	ND	ND	.006
Iron	1.3	329	50.3
Manganese	ND	ND	20.1
Selenium	ND	ND	ND
Silver	ND	ND	ND
Vanadium	ND	ND	0.602
Zinc	ND	27.4	13.7
Toxic trace elements			
Aluminum	18.2	2,290	4,339.4
Arsenic	ND	ND	2.2
Barium	ND	ND	ND
Cadmium	ND	ND	ND
Lead	ND	ND	0.006
Mercury	ND	ND	ND
Other trace elements			
Silicon	146.9	26.2	2.6
Strontium	0.2	ND	78,297.8
Sulfur	98.1	ND	ND
Tin	0.002	ND	0.09
Fluoride	ND	ND	ND

The supplement does not provide any significant source of these "ND" minerals. Do they count as being present if they cannot even be measured? Ninety-two elements, is, well, to put it mildly, a slight exaggeration.

Another popular colloidal mineral supplement contains so

few minerals, and in such minimal quantities, it contributes little to meeting the USRDA*. For the product described in Table XIII, the recommended dose is 1 ounce daily. See how little 1 ounce daily contributes to mineral nutrition.

Table XIII. Colloidal Mineral Content.

Mineral	Mineral Content in Daily Dosage (1 oz)	Percentage of USRDA
Magnesium	100 mg	25
Calcium	25 mg	2.5
Potassium	10 mg	**
Zinc	10 mg	66.7
Manganese	5 mg	**
Silica	1000 ug	**
Selenium	50 ug	**
Chromium	50 ug	**

*USRDA is the US Recommended Dietary Allowance. This is the RDA for adults, male or female, whichever is greater.
** No RDA or USRDA has been established for this nutrient.

In this example, zinc is the only mineral which is present in any significant quantity, suggesting that its nutrient density is quite low. There are many foods which are more nutritious, and less expensive. (One month's supply of this supplement costs approximately $30, or $365 per year!)

...Too Much

On the other hand, some colloidal mineral products contain significant quantities of potentially harmful and toxic elements.

- Sodium
  Some colloidal mineral supplements contain significant amounts of sodium. The manufacturer of Product X contains 20 mg sodium per milliliter. If the recommended dosage of this supplement is 30 milliliters per day, then one obtains an additional <u>600 mg</u> of sodium! The estimated minimum requirement for sodium which is 500 mg. Excess sodium can increase the risks of high blood pressure and osteoporosis.

- Aluminum
  Products Y and Z contain excessive quantities of aluminum. (See Table XII.) Aluminum silicate one of the most common minerals of clay. When individuals with Alzheimer's disease are autopsied, high concentrations of aluminum are found in the brain. This finding certainly does not prove that aluminum causes Alzheimer's disease, but suggests that there may be alterations in aluminum metabolism in individuals with this condition. Some colloidal mineral supplements contain between 1800 and 4400 ppm of aluminum. In contrast, food contains about 10 ppm of aluminum, and furthermore, this aluminum is poorly absorbed.

- Lead
  Its lead quantity is small, but nevertheless, Product Z contains detectable amounts of this toxic metal. (See Table XII.) Any amount of lead is dangerous, since exposure during a lifetime is cumulative. In children, lead interferes with brain development and research has shown that as lead levels in the bloodstream increase, IQ scores decrease.

- Strontium
  The metabolic effects of this metal are unclear. Strontium can accumulate in the body over a lifetime, which increases the likelihood it can be toxic.

What about superior bioavailability? Consider the following manufacturer claims:

- *"Superior absorption in the human intestine."*
  Minerals are absorbed in the free, or elemental, form. This is not dependent upon whether these minerals are suspended in water. More influential is the body's nutriture of that mineral; in other words, if body stores of a mineral are low, then absorption of that mineral from the intestine will be high. This insures that adequate nutriture will be achieved, particularly in times of need. Conversely, if the nutriture of a particular mineral is high, then absorption of that mineral from the intestine will be low. This minimizes risk of mineral toxicity.

- *"Ninety-five percent absorbed."*

If one of the most influential determinants of mineral bioavailability is mineral body stores, then there is no guarantee that 95% absorption occurs, since mineral nutriture varies among individuals. Many minerals have a much lower average absorption rate (i.e., absorption rate of 30% and 10% for calcium citrate and dietary iron respectively). These relatively lower rates are appropriate to accommodate and reflect individual needs; a 95% absorption rate would be too great to be safe, presenting a toxicity risk.

- *"Colloidal minerals are negatively charged, hence increase intestinal tract absorption."*
  Now wait, the wall of the intestinal tract negatively charged. Don't like charges repel one other (the reverse of "opposites attract")? If colloidal minerals are negatively charged, then they would actually be repelled away from the wall of the intestine and travel through the intestine unabsorbed to be excreted with other unabsorbed materials.

The Final Verdict.

How do colloidal mineral supplements measure up? The element content of colloidal mineral supplements varies significantly among manufacturers. Because the source of the elements is clay or earthen material, the consumer is exposed to impurities and other elements which may be toxic and unsafe. Minerals are essential nutrients and must be consumed, but preferably through food or purer supplement forms, as these are not likely to contain large and isolated amounts of unsafe metals.

Solid, Liquid, Gas: Getting Your Minerals.

Did you know that we obtain minerals from our food (solid), drink (liquid), and air (gas)? When researchers want to study the metabolic effects of mineral deficiency, minerals are removed from all food and liquid, and air is filtered to remove airborne minerals. Of course, when it comes to supplements, minerals are available in solid and liquid form. Is one form better than the other?

Liquid supplements are marketed as being more bioavailable than solid pills, but are they? Consider the following:

- ❑ On average, it takes 48 hours (2 days!) for an item to move entirely through the digestive tract (from the mouth to the other end), that is, presuming that it is not absorbed before it is excreted. Supplement tablets dissolve well before 48 hours. Read on to see how you can test whether your supplement dissolves within a reasonable period of time. If you want to have some good, clean fun in the kitchen, turn the page and read the box entitled, "Some Kitchen Chemistry".

Hot Tip:
Some Kitchen Chemistry

Ingredients:
8 oz water
2 tbsp vinegar (acetic acid)
1 spoon
Supplement pill

Directions:

1. Mix water and vinegar.
2. Drop supplement pill into mixture.
3. Allow mixture to sit for 30 to 45 minutes, stirring briefly with spoon every 10 minutes.
4. Check to see if supplement is dissolved.

Principle:
So what have we done? The water-vinegar mixture creates an environment which is similar to the acidic conditions of the stomach. The stirring mimics the churning and muscular contractions of the stomach which aid with mixing and mechanical digestion of food. The stomach begins to empty its contents into the intestine after approximately 45 minutes, the region of the digestive tract wherein most nutrients are absorbed.

Most solid supplements dissolve into a liquid within 30 to 45 minutes of ingestion. To determine whether your supplement tablet meets this criterion, look for the term, "U.S. Pharmacopoeia" (USP).

Supplemental Minerals.

Understanding Supplements.

Let's first start with a general discussion of supplements, including a review of the types and forms available.

Natural or Synthetic?

With regard to dietary supplements, the term "natural" is not well defined. "Natural" implies that the nutrient is extracted from a food source, a process which typically involves exposure of the nutrient source to several chemical solvents. On the other hand, the extraction is a synthetic (man-made) process which occurs in a laboratory! Does this make the supplement natural or synthetic?

> **Hot Tip: US Pharmacoepia: What Does It Mean?**
>
> If a supplement label bears the term "U.S. Pharmacoepia" (USP), then it satisfies purity standards.

Some "natural" supplements may be more harmful than synthetic supplements. For instance, calcium supplements are available in many different forms, including:

- "Natural" sources: Dolomite (bone meal)
  Oyster shell
- Synthetic sources: Calcium carbonate
  Calcium citrate

Dolomite, can be impure and contain high concentrations of lead and other toxic metals. The metal impurities depend upon the animal's exposure. In this case, the natural form can be more harmful than its synthetic form.

Therapeutic Supplements.

To act as nutraceuticals, some minerals must be consumed in amounts which exceed the RDA. Remember adequate vs. optimal? (or if a little is good, then more must be....better.) But also remember secondary deficiencies and nutrient interaction (or if a little is good, then more must be...worse.) Above all, minerals, when used as nutraceuticals, must be considered separately and in the context of the individual's needs, age, and diet. Part VI entitled, "Minerals As Nutraceuticals" addresses minerals separately and their possible usefulness.

Selenium and chromium are among today's popular minerals because of their putative health benefits. Interest in selenium has emerged because of its indirect role as an antioxidant. Interest in chromium abounds because of its role in carbohydrate metabolism. Let's examine each of these minerals and their supplements more closely.

- Selenium mineral
  Selenomethionine supplement
  "Natural" selenium is manufactured in the laboratory by adding selenium to a yeast culture. The yeast cells absorb the selenium and convert it to its natural form, selenomethionine (a selenium-methionine complex); methionine is one of the essential amino acids. The selenomethionine complex is the "natural" form of selenium as produced by yeast, but this is not the same form of selenium as found in the human body, or the biologically active form which participates in antioxidation. Some selenium supplements which are sold as "natural" may merely consist of selenium mixed with yeast.

- Chromium mineral
  Glucose tolerance factor supplement
  In the human body, chromium is most studied as part of the glucose tolerance factor (GTF). The GTF is a complex found on the surface of cells. It consists of several components, including: chromium, nicotinic acid (a form of vitamin B_3), and the amino acids: glycine, glutamate, cysteine, or glutathione. Like selenomethionine, GTF can be assembled by yeast if chromium is added to the yeast culture. The GTF enhances the action of insulin, which is integral for transporting glucose into cells for metabolism and controlling blood sugar levels. In fact, abnormally elevated blood sugar, or hyperglycemia, is a symptom of chromium deficiency.* Chromium supplements are commonly available as chromium picolinate (a mineral-chelator complex) or as chromium-GTF. Most individuals

can assemble chromium and other necessary components to form GTF. New research has suggested, however, that some individuals utilize chromium poorly and need to have the preformed GTF. There has been some limited research which suggests that GTF may be helpful in treating non-insulin dependent diabetes (or type II diabetes), which is caused by an insensitivity to insulin.

* Elevated blood sugar may be caused by a number of conditions in addition to chromium deficiency. If you have hyperglycemia, consult your physician to identify the cause, and obtain the appropriate treatment for this condition.

The Waiting Game: Time-Released Supplements.

The bioavailability of a mineral is very dependent upon the body's nutriture of that mineral. In other words, if the level of a particular mineral in the bloodstream or if tissue storage is low, then the intestine is more efficient at absorbing the needed mineral. Conversely, if the level of a mineral in the bloodstream or if tissue storage is high, then the intestine is less efficient at absorbing that mineral. Hence, supplements, particularly those which deliver megadose amounts (i.e., quantities at least 10 times the RDA) can cause the level in the bloodstream to rise abruptly. Consequently, further absorption of those nutrients is depressed.

Time-released supplements are intended to prevent the abrupt and rapid rise in nutrients by dissolving slowly in the intestine. Nutrients, particularly minerals, are absorbed at very specific sites along the intestine. Once the mineral passes by the absorption site, it is less likely to be absorbed further down in the intestine as the pill is dissolved. In reality, nutrients from time-released preparations are often poorly absorbed.

CHAPTER FIVE
What Do You Look Like?

Are You at Risk?

Is There a Senior Discount with That?...The Elderly

The elderly have unique needs and greater demands for minerals, yet, paradoxically, the nutritional requirements of the elderly have not been well researched. The elderly are at risk for a number of mineral inadequacies and deficiencies because of certain characteristics.

- "Eating like a bird."
  (Actually, birds can eat up to twice their weight in food each day, but you know what we mean.) The elderly consume less food, and therefore, fewer essential nutrients, including minerals. Physical limitation, social isolation, economic restriction, or combinations of these conditions increase this likelihood of reduced consumption.

- "The tea and toast syndrome."
  "Tea and toasters" are usually elderly women. These individuals usually eat a diet containing food low in nutrient density (tea), high in carbohydrates (toast), and low in protein (meat).

- "Poor bioavailability."
  There are two reasons for reduced nutrient bioavailability in senior citizens:

  - Low production of stomach acid and digestive enzymes which aid in nutrient absorption.
  - Drug-nutrient action.
    Seniors account for the highest sales of prescription and nonprescription medications (more than 80% of older adults use more than two medications daily and 61% take nonprescription medications), many of which may interfere with nutrient absorption. Refer to the section below titled, "Pills, Pills, and More Pills" for more information about the effect of specific medications on mineral nutriture.

There are certain "problem" minerals in the US population; these minerals are often inadequate in the American diet. In general, nutriture of these minerals is poorer among the elderly than younger adults. During times of physical stress, the elderly individual is at even greater risk for a severe deficiency because mineral stores are low and unavailable in times of need.

- ❑ Calcium
  In both men and women, calcium absorption decreases with age. This is due, in part, to:

  - ❖ A reduction in stomach acid production. Stomach acid helps maintain calcium in specific chemical state to enable its absorption.

  - ❖ Less exposure to sunlight. Yes, less exposure to sunlight. A compound in the skin is converted to vitamin D when exposed to the sun's ultraviolet rays. In fact, a mere 15 minutes of casual sun exposure of face, hands, and arms is sufficient to provide the RDA for vitamin D. Vitamin D enhances calcium's bioavailability. Dairy products are fortified with vitamin D which enhances the absorption of its calcium. Calcium supplements that contain vitamin D are more bioavailable. Refer to Table VI for a listing of calcium-rich foods.

- ❑ Iron
  Iron deficiency in the elderly may be due to the following:
  - ❖ Poor absorption of iron due to decreased stomach acid production.
  - ❖ Inadequate dietary intake;
  - ❖ Blood loss due to chronic disease such as hemorrhoids or ulcers;

  The RDA of 10 mg is adequate for the elderly. Refer to Table VII for a listing of iron-rich foods.

- ❑ Zinc
  - ❖ Research has shown that the elderly consumes inadequate levels of zinc. Absorption of zinc declines with age. Refer to Table XIV for a listing of zinc-rich foods.

Table XIV. Zinc-Rich Foods.

Zinc-Rich Food	Zinc Content (mg/serving)	Serving Size	Calories per Serving
Oyster, raw	32	3 oz	50
Crab, Alaska King, cooked, moist heat	6.5	3 oz	83
Cereal, All-bran	11	1 cup	212
Garden burger (meatless burger)	7.5	1 each	140
Lobster, Spiny, cooked, moist heat	6.2	3 oz	122

Pills, Pills, and More Pills...Chronic Medications.

This section provides you with general information about chronic medications and their effect upon mineral nutriture. Within each category, there are numerous medications and brands which may have specific and varying effects upon mineral nutrition. Please consult your health care professional for information about your specific medication.

Antacids.

Antacids alleviate heartburn and stomach pain from excess acid. There are various types of antacids, each with different effects upon mineral nutriture because of their content. Types and nutritional effects of buffering antacids (drugs which neutralize stomach acid) are described below.

- Aluminum hydroxide antacid
  - Aluminum toxicity
  - Decreased bioavailability of phosphorus and calcium

- Magnesium hydroxide antacid
  - Magnesium toxicity
  - Decreased bioavailability of phosphorus and calcium

- Sodium bicarbonate antacid
  - Excess sodium intake
  - Decreased bioavailability of calcium and iron

There are also "antacids" which actually reduce the production of stomach acid. These so-called "H_2 blockers" reduce calcium and iron absorption because absorption of these minerals is enhanced by hydrochloric acid.

Antidepressants.

Many antidepressants have an "anorectic" effect. Loss of appetite is induced by side effects such as nausea, vomiting, dry mouth, diarrhea, reduced salivation, and stomach upset.

Arthritis Medications.

D-penicillamine is a common prescription medicine for arthritis. This medication has several effects upon nutrients, including an anorectic effect, primarily because of the same side effects as listed above. With regard to mineral nutriture, D-penicillamine can bind to iron, zinc, and other minerals. Hence, this mediation should be taken on an empty stomach (i.e., 1 hour before or 2 hours after a meal), and at least 1 hour apart from any other medication, food, or beverage (with the exception of water).

Coronary Heart Disease Medications.

There are several types of coronary heart disease medications. This section will focus upon medicines which lower blood cholesterol and triglyceride (another type of fat found in the bloodstream which, when elevated, can increase coronary heart disease risk), by interfering with the absorption of bile and fat from the intestine. These medications are collectively referred to as "bile acid sequestrants."

Here's how sequestrants work to reduce cholesterol. Reduced bile absorption from the intestine means that more bile is excreted from the colon. As more bile is excreted, the body needs to replenish its supply. Cholesterol is needed to make the additional bile. The cholesterol comes from cholesterol supplies in the blood stream, thereby reducing the level of blood cholesterol.

Reduced fat absorption from the intestine means that less fat enters the body's cells. The lower amounts of cellular fat reduce fat's ability to exert its detrimental effects on the body (i.e. weight gain, and increased production of cholesterol by the body).

Despite our "fear of fat," scientists recognize that a little dietary fat is beneficial because it acts as a vehicle to bring healthy fatty components of the diet into the bloodstream. Healthy fatty components--what are they? What about fat-soluble vitamins such as vitamins A, D, E and K? Remember that vitamin D enhances calcium bioavailability? Individuals consuming bile acid sequestrant medications for long periods of time have an increased risk of osteomalacia, a bone disease related to vitamin D deficiency.

The old adage, "Throwing the baby (fat-soluble vitamins) out with the bath water (bile)" rings true for these medications. Individuals taking these types of medication for long periods should take a supplement which contains the fat soluble vitamins A, D, E and K.

Oral Contraceptives.

Some women taking oral contraceptives exhibit low levels of iron and zinc in the bloodstream. It is still not clear whether this phenomenon is strictly due to the chronic medication or merely reflects the nutrient-poor diet among many young women. Perhaps additional variables account for poor iron and zinc nutriture in these women, for instance, length of time on medication, specific oral contraceptive used, and individual response to the medication.

Hypertensive Medications.

Several types of medications lower blood pressure. This section will focus upon antihypertensive medications that increase the excretion of water, sodium, and chloride. Medications with this action are collectively referred to as "diuretics." Unless diuretics are also classified as "potassium-sparing diuretics," they will also increase urinary potassium loss. Because potassium depletion can cause irregular heartbeat, it is recommended that individuals taking non-potassium-sparing diuretics consume a potassium supplement (this must be approved by a physician) and/or potassium-rich foods, such as those listed in Table XV below.

Table XV. Potassium-Rich Foods.

Potassium-Rich Food	Potassium Content (mg/serving)	Serving Size	Calories per Serving
Bamboo shoots, fresh	402	0.5 cup	20
Spinach, fresh	312	1 cup	12
Celery, fresh	172	0.5 cup	10
Lettuce, Romaine, fresh	162	1 cup	9
Lettuce, butterhead	144	1 cup	7
Radish, fresh	140	0.5 cup	7
Coffee, brewed	96	6 oz	4
Tea, brewed	88	8 oz	2
Squash, zucchini, fresh	50	1 medium	2

Hot Tip:
Potassium & Food Labels

The Food & Drug Administration (FDA) describes the potassium content of food according to the following definitions:

- "High-potassium": 700 mg or more per serving.
- "Good source of potassium:" 350 to 665 mg per serving.
- "More or added potassium": at least 350 mg more per serving than standard recipe.

Laxatives.

Some laxatives increase the rate of movement of food and nutrients through the intestine. Consequently, essential nutrients such as vitamins and minerals not well absorbed. Stool softeners such as mineral oil can bind to fat-soluble vitamins which are subsequently eliminated with the mineral oil. If the bioavailability of vitamin D is reduced, then calcium absorption is also decreased.

Weekend Warriors…Athletes.

Are athletes unique in their nutritional needs? Given the plethora of dietary supplements intended to optimize performance, maximize muscle growth, and enhance fat-burning, it would appear so. Curiously, there is consumer demand for these substances, but research does not support a special need for supplements. As a group, athletes in search of the "edge" are, with the exception of dieters, most vulnerable to exaggerated nutritional and supplement claims.

With regard to minerals, let's address two issues relevant to the athlete.

- **Sports Anemia**

  Whole blood is comprised of two fractions, the aqueous part (the plasma or watery phase) and the cellular part (which consists of red blood cells and immune cells). When the percentage of red blood cells in whole blood is measured, this is called the hematocrit. Hematocrit measures are commonly used to diagnose iron deficiency anemia, because the number, and therefore, the percentage of red blood cells declines if iron is not present. But what were to happen if the number of red blood cells remained the same, but the volume of the aqueous phase were to increase? This would result in an artificial lowering of hematocrit, independent of iron nutriture. During exercise, the plasma volume increases, probably as an adaptive response to enable a better flow of blood to muscles. Because this phenomenon results in a lower hematocrit, there is a mistaken belief that athletes have a higher requirement for iron or that many athletes are iron deficient. To determine with certainty whether a low hematocrit is due to iron deficiency, or due to an increase in plasma volume, other blood chemicals must be measured.

- **Electrolyte Losses in Sweat**

  The average athlete loses more water, relative to electrolytes (i.e., sodium, chloride, potassium, magnesium) through sweat so that the body actually becomes more concentrated in electrolytes! Hence, there is no pressing concern for electrolyte replacement. Instead, the priority becomes replacement of fluid, since a significantly large proportion of water is lost. So, what about the athletic replacement drinks? Typically, these have three important nutritional characteristics.

- Fluid content.
  This is the most important role of an athletic replacement drink. Lack of fluid replacement can result in over heating and fatigue.
- Electrolyte content.
  There is no physiological necessity for electrolyte replacement in the average athlete. Furthermore, the average athlete consumes more food to compensate for increased energy needs. The additional food (if nutrient-dense and healthy) provides the quantity of minerals necessary to replace losses incurred during exercise.
- Carbohydrate content.
  Carbohydrate provides necessary fuel for the working muscle during exercise. Fluid containing carbohydrates, when consumed during exercise maintain blood sugar levels, so that a readily available source of fuel is available for the working muscle.

No Meat for Me, Thank You...Vegetarian Diets.

More individuals are adopting vegetarian diets as a means to maximize heath and obtain optimal nutrition. The "vegetarian" diet actually encompasses several types of diets, with various degrees of restriction. See Table XVI to understand the differences among vegetarian diets.

Table XVI. Types of Vegetarian Diets.

Vegetarian Diet	Foods Consumed
Vegan	Plant products only
Lactovegetarian	Plant products plus dairy products
Ovolactovegetarian	Plant products plus dairy products and eggs
Fruitarian	Fruits, nuts, honey, and vegetable oils

As a rule of thumb, the more restrictive the diet, the more

difficult it is to obtain adequate nutrition. Since animal products supply a large proportion of calcium, iron, and zinc, one should pay special attention to obtaining adequate nutriture of these minerals.

- Calcium
  Dairy products can supply as much as 75% of the daily need for calcium. If animal products, including dairy product are eliminated from the diet, then calcium-containing plant foods must be consumed to meet the calcium requirement. Table XVII lists calcium-rich plant foods.

Table XVII. Calcium-Rich Plant Foods.

Calcium-Rich Plant Food	Calcium Content (mg/serving)	Serving Size	Calories per Serving
Almonds	166	0.5 cup	418
Kale	158	1 cup	7
Navy beans	95	1 cup	200
Great Northern beans	90	1 cup	
Soybeans, boiled	88	0.5 cup	149
Soy milk	60	1 cup	100
Blackeyed peas	22	0.5 cup	100
Broccoli fresh	42	1 cup	24
Tofu, regular	30	1 oz	22

- Iron
  Plant foods contain a certain type of iron, nonheme iron. The bad news first....nonheme iron is less bioavailable and its absorption is further reduced by phytic acid (found in unleavened grain products), oxalic acids (common in spinach, chard and beet greens), dietary fiber (contained in whole grains) and tannic acid (found in some teas). The good news now...to facilitate absorption of iron, foods high in iron and vitamin C (ascorbic acid) should be consumed in the same meal. Tables XVIII and XIX describe iron-rich plant foods and vitamin C-rich plant foods which can be combined to enhance iron absorption.

Table XVIII. Iron-Rich Plant Foods.

Iron-Rich Plant Food	Iron Content (mg/serving)	Serving Size	Calories per Serving
Sunflower seed kernels	4.2	0.5 cup	372
Spinach, cooked	4	1 cup	42
Black beans, cooked	4	0.5 cup	114
Almonds	3.5	0.5 cup	418
Pinto beans, cooked	3.2	0.5 cup	94
Raisins	2.7	0.5 cup	218
Navy beans, boiled	2.5	0.5 cup	98
Soy beans, boiled	2.5	0.5 cup	149
Prunes	2.3	0.5 cup	224

Table XIX. Vitamin C-Rich Plant Foods.

Vitamin C-Rich Plant Food	Vitamin C Content (mg/serving)	Serving Size	Calories per Serving
Papayas, fresh	188	1 medium	119
Peppers, sweet, red, fresh	141	1 medium	20
Peppers, hot chili, green, fresh	109	1 each	18
Kiwifruit, fresh	74	1 medium	46
Orange, fresh	70	1 medium	62
Peppers, sweet, green, fresh	66	1 medium	20
Broccoli, cooked	58	0.5 cup	22
Strawberries, fresh	42	0.5 cup	22
Cabbage, fresh	36	1 cup	17

- Zinc

  Animal products such as lean meats, oysters, fish, milk, and egg yolks are excellent sources of zinc. By eliminating these foods from the diet, the vegan also excludes zinc sources. Whole grains and cereals can provide zinc, but the usually high dietary fiber content of the diet can interfere with the absorption of this mineral; therefore, the vegan must insure frequent daily servings of these foods. Zinc-rich plant foods may insure adequate zinc nutriture in the vegan.

See Table XX for a description of zinc-rich plant foods.

Table XX. Zinc-Rich Plant Foods.

Zinc-Rich Food	Zinc Content (mg/serving)	Serving Size	Calories per Serving
Hamburger patty, meatless	7.5	1 each	140
Garden burger (meatless burger)	7.5	1 each	140
Cereal, 40% bran flakes	5.2	1 cup	127
Cereals, barley	5.4	1 cup	153
Cereals, rye	5.3	1 cup	144
Cereals, wheat	5.8	1 cup	158
Blackeyed peas	3.4	0.5 cup	100
Peanut butter	1.0	2 tbsp	190

Lose 10 Pounds in 2 Days...The Dieter.

Dieting is one of the most common pastimes in America! At any given time, approximately 25% of men and 50% of women are attempting to lose weight. Unfortunately, many of the latest-and-greatest diets are not healthy, sound, or sensible. Severe calorie restriction (i.e., 1000 calories per day or less), or the elimination or restriction of particular food groups makes many weight loss diets unsafe and unsuccessful. A current and popular weight loss diet promotes reduction of carbohydrate foods, an increase in protein, and a slight increase in fat. This book is not intended to address weight loss in detail, but two critical questions that should be addressed when adopting a dietary regimen are:

Are You Adequate?

What about the micronutrient (vitamin and mineral) content of the diet? Does the diet provide adequate vitamin and mineral nutriture? Severe calorie restriction of 1000 calories per day or fewer does not provide adequate nutrition. In other words, there is simply not enough food to supply vitamins and minerals in the quantities necessary to meet biological needs. Typically, this level of calorie intake is too severe to support healthy weight loss anyway.

Are You Balanced?

What about the macronutrient (carbohydrate, protein, and fat) content of the diet? Does the diet provide an appropriate proportion of macronutrients, giving a balanced intake? Current recommendations for a healthy weight-loss diet include:

Carbohydrate	55 to 60% of calories
Protein	15 to 20% of calories
Fat	20 to 25% of calories

To date, no health professional organization has promoted or endorsed a 40/30/30 (carbohydrate/protein/fat) weight loss diet. There is limited research about the effectiveness and long-term weight maintenance on this diet. Health care professionals do not back this recommendation because they are still saying, "Show me the research!"

CHAPTER SIX

Minerals as Nutraceuticals

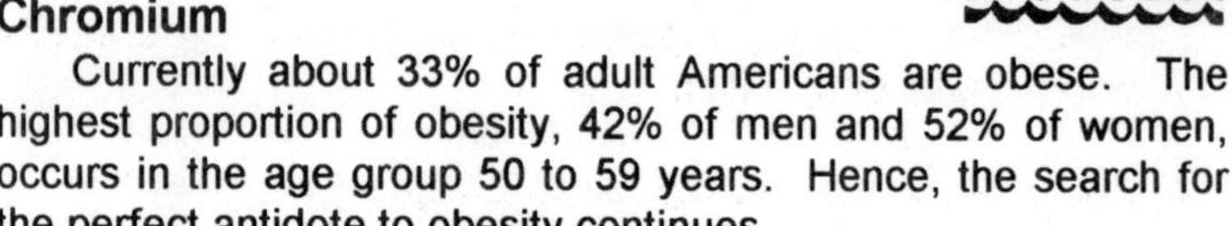

Fat Burner?

Chromium

Currently about 33% of adult Americans are obese. The highest proportion of obesity, 42% of men and 52% of women, occurs in the age group 50 to 59 years. Hence, the search for the perfect antidote to obesity continues.

Does chromium help burn fat? To answer the $64,000 question of the 1990's, let's take a look at the scientific evidence. Most of the research has shown that chromium supplementation does not have a beneficial effect upon body composition. In other words, chromium supplementation does not cause a reduction in body fat and/or an increase in lean body mass. Chromium, at various dosage levels and under various conditions of exercise, has no effect on the body composition of individuals, including such individuals as exercising men, athletes in general, and the obese. Consider a sampling of the latest research examining the effect of chromium supplementation:

- *A study was conducted to see if chromium supplementation would enhance the beneficial effects of strength training.* For eight days, a group of men received either placebo of chromium chloride or chromium picolinate, at a dose of 172 to 185 ug per day. These men also engaged in routine weight training and exhibited improvements in muscular strength, but chromium supplementation was not responsible for the effect. We know because exercising individuals receiving the placebo exhibited the same improvement in strength and body composition as those exercisers receiving chromium supplementation. Researchers concluded that the only variable which significantly increased muscular strength was exercise! (Big surprise!)

- *Research was performed to study the retention of chromium in the body and its effect upon performance in athletes.*

Football players were given either placebo or chromium picolinate supplementation (200 ug per day). Subjects receiving chromium supplements demonstrated a five-fold increase in chromium excretion in the urine, suggesting that these individuals were producing very expensive urine! But seriously, this suggests that chromium supplements are not retained in the body, and that 200 ug per day does not have a significant effect upon the body's physiology. As expected, chromium supplementation did not improve body composition or strength.

- *An experiment was designed to test whether chromium supplementation could be used as an aid for weight loss.* Obese, active-duty Navy personnel were given either a placebo or 400 ug of chromium picolinate (twice the manufacturer's usual recommended dosage) and enrolled in a physical conditioning program (3 times per week, 30 minutes per session, 16 weeks duration). There was no difference between the placebo and chromium-supplemented group. Needless to say, the Navy is not adopting chromium picolinate to improve performance in the military!

If the research does not support chromium, then how did the idea of "chromium as a fat burner" arise? Let's look at metabolism at the molecular level. There is an adage among biochemists: "Fat burns in the fuel of carbohydrate." Translation for non-biochemists: We need carbohydrate to burn fat. In metabolism, many chemicals interact to produce a final outcome. Carbohydrate interacts with fat to produce a final outcome of fat oxidation (burning). All this activity is occurring inside the cell. Thus, fat is burned inside the cell, and carbohydrate is aiding this intracell process. Obviously, before the wheels can be set in motion, carbohydrate (typically as glucose) must enter into the cell. One of the compounds which allows glucose to enter into the cell is the "glucose tolerance factor" (GTF). And guess what mineral is part of GTF--chromium. But GTF also consists of several other components, including: nicotinic acid (a chemical relative of vitamin B_3), and the amino acids glycine, glutamate, cysteine, or glutathione. In addition, fat needs to enter the cell before it can be burned. To put so much faith in one mineral which is an important, but not the only essential component of fat oxidation, is metabolically short-sighted and over-simplified. Could this be why research has not shown that chromium supplementation decreases body fat? What works on paper may not work in the complex milieu of metabolism.

Glucose Metabolizer?

Chromium

Despite disappointing results as discussed in the previous section, there is renewed interest in chromium as a nutraceutical for the prevention and treatment of high blood glucose and insulin insensitivity. Hyperglycemia (i.e., an excessive level of glucose in the bloodstream), high insulin levels, and insulin insensitivity are characteristics of non-insulin dependent diabetes mellitus (NIDDM), a disease associated with aging, obesity, and genetic predisposition. Because of the "graying of America" and increased incidence of obesity (estimated to be one-third of all adult Americans), the incidence of NIDDM has also risen.

Preliminary findings show that chromium may have a promising role in alleviating hyperglycemia. Consider a sampling of the latest research examining the effect of chromium and its relationship to age and blood sugar levels.

- *A survey was conducted to examine how age affects the level of chromium in the body.*
  As one ages, there is a significant reduction in the level of chromium in hair, sweat, and blood. These data suggest that age-related reductions in chromium may be, in part, responsible for age-related increase in NIDDM, and, as some scientists now speculate, the increase in blood cholesterol and triglyceride levels also associated with advanced age.

- *A research project was developed to study the effect of chromium upon insulin levels.*
  Healthy, non-obese subjects were given either a placebo or chromium nicotinate (containing 220 ug of chromium). Chromium lowered insulin levels in individuals who tended to exhibit higher insulin levels. Note that higher insulin levels are thought to be a risk factor and characteristic of NIDDM.

- *An experiment was designed to examine whether glucose tolerance factor (GTF) alleviates symptoms in diabetic rats.* When GTF was injected into diabetic rats, within two hours there was a significant reduction in blood sugar and fats, suggesting that GTF, or perhaps the chromium component of GTF is important as an anti-diabetic chemical.

Based on these and other research, chromium seems to be a promising nutraceutical for the alleviation of NIDDM.

Table XXI. Chromium-Rich Foods.

Chromium-Rich Food	Chromium Content (ug/serving)	Serving Size	Calories per Serving
Yeast, brewer's	32	1 oz	69
Pepper, green	22	4 oz	26
Apple	20	1 medium	83
Banana	17	6 oz	156
Cornmeal	12	3.5 oz	43
Spinach, fresh	11	4 oz	25
Bread, whole wheat	12	1 oz	69
Wheat bran	11	1 oz	61
Bread, rye	8.6	1 oz	94

Free Radical Terminator?
Selenium

Remember that many minerals serve as cofactors, or compounds necessary for enzyme function and metabolism. Selenium is a cofactor for the enzyme which prevents the formation of "lipid peroxides." This selenium-dependent enzyme is called glutathione peroxidase. What is the significance of this? Lipid peroxides are formed by free radicals and can damage genetic material such as DNA (increasing cancer risk), the walls of the artery (creating a coronary heart disease risk), and cells of the eye (causing cataracts and macular degeneration). Because of the many diseases associated with free radical damage, there is tremendous interest in preventing the formation of free radicals, such as through selenium supplementation. Where are we today, and can selenium be thought of as an effective antioxidant, akin to beta-carotene and vitamins A, C, and E?

Selenium vs. Free Radicals.

Consider the following in current research in the relationship between oxidative stress (a measure of potential free radical damage) and selenium supplementation.

- *Scientists examined whether selenium supplementation could provide antioxidant defense against free radical damage in the elderly.*
  When nursing home residents were given supplements containing selenium (containing a daily dose of 100 ug selenium), the activity of the antioxidant enzyme, glutathione peroxidase increased. This suggests that the selenium-dependent enzyme was not functioning to maximal capacity in these individuals; selenium supplementation allowed the activity of this enzyme to increase to a pre-established maximum level. Hence, the ability to defend against free radical damage was increased because the activity of this enzyme was also increased.

Selenium vs. Cancer.

Cancer is the second leading cause of death for American adults and is projected to be the number one cause of early death in the next century. Although there are many causes for the several different types of cancer, it is estimated that 60% of all cancer cases are diet-related. Could selenium be an important dietary influence also?

Selenium is an "ironic" mineral; too much can cause cancer...too little can also increase cancer risk. How much is just enough? Selenium has a small "window of safety" (the amount which is safe, adequate, and optimal). An intake of 200 ug selenium is considered to be within the window of safety. Because selenium plays a role in the antioxidant defense system (along with beta-carotene and vitamins A, C, and E), it has an obvious role in cancer risk reduction. The research studies described below demonstrates that several provocative findings:

- The chemical form of selenium influences its biological activity; and
- Selenium compounds can reduce the progressive stages of cancer.

- *An experiment was undertaken to examine the most effective form of selenium and how it worked to reduce cancer risk.*
  It has long been suspected that sodium selenite is a more toxic form of selenium. (Remember the small window of safety?) Some "organoselenium" compounds (selenium attached to organic chelators) may be less toxic, and in fact, more biologically effective at reducing cancer risk. These researchers discovered that of all the organoselenium compounds they tested, (take a deep breath now) 1,4-phenylenebismethylene selenocyanate was most effective at reducing breast tumor development in rats. For simplicity, lets call this chemical "p-XSC." How does it work? Some carcinogens bind to DNA, causing abnormalities in the genetic machinery. These "DNA adducts" cause a normal cell to become abnormal, then cancerous. p-XSC not only prevents DNA adducts from being formed, but also inhibits cancer progression by inhibiting an enzyme necessary for tumor growth.

Selenium vs. Coronary Heart Disease.

Coronary heart disease is the major killer of Americans, both men and women. Each year approximately 500,000 individuals die of heart disease in the US. There are several processes involved in the development of atherosclerosis, or coronary heart disease. Free radical and lipid peroxide damage of the artery lining is thought to be a common cause of heart disease. Given selenium's role in preventing lipid peroxidation, we suspect that selenium is an important nutraceutical in the prevention of atherosclerosis. Read on for more evidence of this relationship.

- *A research project was conducted to determine why the population living in Northern Finland had less coronary heart disease than individuals in other parts of Finland.*
  Samis, or the Northern Finlanders examined in this study, had significantly higher levels of biochemicals which <u>decrease</u> coronary heart disease (i.e., vitamin E and selenium). Paradoxically, they also had higher levels of

biochemicals which increase coronary heart disease risk (i.e., total cholesterol and LDL or "bad" cholesterol). So, it appears vitamin E and selenium protect the Samis from heart disease, despite their high cholesterol levels!

Selenium vs. Cataracts.

Researchers speculate that retinal and corneal damage resulting from ultraviolet and short-wavelength light, may damage these cells, making them vulnerable to free radical damage. Antioxidants such as vitamin C and carotenoids seem to decrease the risk of free radical damage. Populations which consume more fruits and vegetables (common sources of antioxidant vitamins and phytochemicals) exhibit lower incidence of macular degeneration and cataracts.

Because of the role of selenium in antioxidation, could this mineral also serve to protect the eye from the loss of visual ability that occurs with aging?

Individuals with low levels of antioxidants in the bloodstream have a greater chance of developing cataracts.

A survey of the selenium content in individuals with cataracts is described below...

Individuals with a cataract have lower selenium levels in the blood, lens, and aqueous humour (the clear, jelly-like covering of the eye). Scientists believe that this reflects poor antioxidant defense systems in these individuals, which leads to cataract formation.

Table XXII. Selenium-Rich Foods.

Selenium-Rich Food	Selenium Content (ug/serving)	Serving Size	Calories per Serving
Red Swiss chard	65	4 oz	22
Oats	48	3 oz	51
Orange juice	33	6 oz	77
Rice, brown	33	3 oz	94
Wheat germ	32	1 oz	103
Brazil nuts	29	1 oz	186
Barley	21	3 oz	301
Bread, whole wheat	19	1 oz	70
Wheat bran	18	1 oz	61
Garlic	7	1 oz	42

Where is this research on nutraceuticals taking us? Agriculturalists and food manufacturers are now interested in creating "pharmafoods" or "functional foods." These are foods which have medicinal or purposeful function in health. For instance, when garlic is fertilized with selenite, it becomes enriched in selenium. Fortunately, the garlic converts the selenium to a less toxic organoselenium compound called selenium methylselenocysteine, which is basically selenium attached to the amino acid, cysteine. Case in point: Between 1972 and 1992, the population in Finland has experienced a decline in deaths due to coronary heart disease (55% decline among men and 68% decrease among women!). Several dietary and lifestyle changes are responsible (i.e., reduction of total fat, saturated fat, and cholesterol; increased fruit and vegetable intakes), including a three-fold increase (as of 1985) in the use of fertilizers containing selenium!

Hypertension Buster?
Calcium, Magnesium, and Potassium

More than 50 million Americans suffer from hypertension. It is estimated that hypertension is responsible for approximately 31,000 deaths each year. Because this condition can cause coronary heart disease, kidney disease, stroke, and poor blood circulation in the legs and sudden death, it must be diagnosed and controlled. Dietary and lifestyle changes are implemented first, before medication, but how effective is dietary change such as sodium restriction?

Sodium restriction is inconsistent in its ability to lower blood pressure in some individuals. In fact, only "sodium sensitive" individuals, or approximately 30% of individuals with hypertension can be helped through dietary sodium restriction. (Sodium sensitive individuals develop hypertension as a result of excessive sodium.) The remaining individuals do not respond to sodium restriction, and blood pressure must be lowered through weight loss and/or medication. Because anti-hypertensive medications can be a life-long treatment, with side effects, there is a search for non-pharmacological treatments for hypertension. Advances show that a combination of calcium, magnesium, and potassium may be promising therapy for this disease. In this case, adequate appear to be equivalent to optimal. Intakes of calcium and magnesium at the Recommended Dietary Allowance (RDA) (See Table XXIV.) and potassium at the Estimated Minimum Requirement (EMR) (500 mg) is enough to decrease blood pressure.

Table XXIII. RDA for Calcium And Magnesium.

Mineral	RDA Male (25 to 50 years)	RDA Female (25 to 50 years)	RDA Male (51+ years)	RDA Female (51+ years)
Calcium (mg)*	800	800	800	800
Magnesium (mg)	350	280	350	280

How does it work? Most studies have examined the metabolic interaction between calcium and sodium. For instance, for every 150 mg of calcium consumed, 20 mg of sodium are excreted in the urine. Calcium may lower blood pressure, by eliminating excess sodium from the body. Read on to examine another study which shows that the opposite is true; that is, that excess sodium causes a net loss of calcium from the body.

❑ *A research project was designed to study calcium metabolism in sodium-sensitive individuals.*

Hypertensive men showed greater urinary loss of calcium and lower levels of calcium in the bloodstream. Furthermore, the degree of salt sensitivity correlated with calcium in the bloodstream. In other words, a higher salt intake resulted in lower calcium concentrations in the bloodstream. So, it may be that sodium intake itself is not a direct cause of hypertension, but rather the calcium losses caused by sodium intake is the culprit.

For information about food sources of minerals discussed in this section, refer the previous tables:

- Table VI. Calcium-Rich Foods
- Table X. Sodium-Rich Foods
- Table XV. Potassium-Rich Foods

Table XXIV. Magnesium-Rich Foods.

Magnesium-Rich Food	Magnesium Content (mg/serving)	Serving Size	Calories per Serving
Cereals, all-bran	318	1 cup	212
Spinach, cooked	78	0.5 cup	21
Potato, baked with skin	55	1 medium	220
Okra, fresh	29	0.5 cup	19
Coffee, brewed	9	6 oz	4
Tea, brewed	7	8 oz	2
Squash, zucchini, baby, fresh	4	1 medium	2

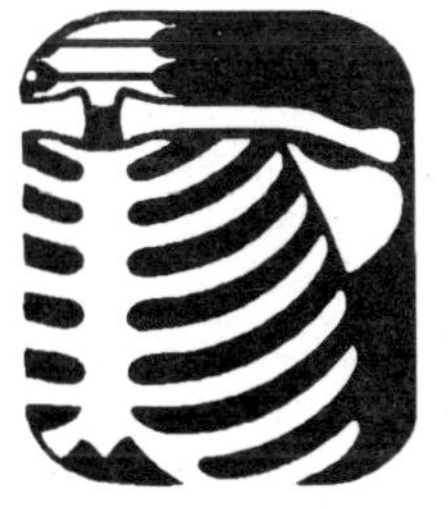

Bone Strengtheners? Boron, Calcium, Fluoride, Magnesium, Phosphorus, and Copper, Manganese and Zinc

Osteoporosis is a major medical problem which afflicts elderly women and men although most of the 25 million Americans affected are women. About 25% of all women over 50 develop osteoporosis. Osteoporosis occurs in older individuals because they have experienced a longer period of bone loss. This disease can not only be debilitating because of bone pain and fractures, it can fatal; 12% to 25% of all elderly person who suffer hip fractures eventually die from fracture-related complications.

It's more than just calcium that does it. New research findings cause us to add minerals to the nutraceutical list for osteoporosis. Listed below is an overview of the minerals that have a positive effect on bone mineral density.

Boron vs. Osteoporosis.

So little is known about boron, that scientists still do not know whether it is an essential nutrient or how it actually affects bone mineral density. Chronically low intakes of boron, along with several other nutrients, may increase osteoporosis risk. Specifically, boron is thought to be involved in the metabolism of the following biochemicals which affect bone mineral density:

- Vitamin D
  This vitamin facilitates calcium retention in the body by:
  - increasing the bioavailability of calcium; and
  - decreasing the urinary excretion of calcium; and
  - increasing the deposition of calcium into bones.

- Estrogen
  This hormone enhances maximum bone density by:
  - increasing the activity of osteoblasts, or cells which deposit calcium into bones; and
  - increasing the bioavailability of calcium.

Good sources of boron include noncitrus fruits, leafy vegetables, nuts, and beans.

Calcium vs. Osteoporosis.

Research has shown unequivocally that adequate calcium intake dietary calcium is essential for maximum bone mineral density. Controversy still rages about the appropriate level of calcium intake. Many believe that current RDAs are insufficient to provide maximum calcium nutriture. Consider the following:

- The calcium RDA (25 to 50 years) is 800 mg; however, the USRDA is 1200 mg, with this higher amount designed to encourage a higher calcium intake.

- Current research is attempting to determine whether the calcium RDA for individuals 11 to 24 years should be increased from 1200 mg to 1500 mg.

Realize that calcium is not the only answer to maximizing and maintaining bone density. In postmenopausal women, in particular, calcium intakes up to 2000 mg per day did not prevent bone loss in the spine, hip, or wrist as successfully as estrogen replacement therapy. Ideally, calcium (recommended dosage 1500 mg per day) with adequate vitamin D (or exposure to sunlight) and estrogen replacement therapy* should be implemented to reduce bone loss and osteoporosis related fractures.

*For some women, there may be contraindications to estrogen

replacement therapy. The decision to begin estrogen replacement therapy should be made by a woman and her physician.

Fluoride vs. Osteoporosis.

Fluoride can form fluoroapatite crystals in bones and teeth, rather than hydroxyapatite. Individuals consuming fluoridated water demonstrate greater bone mineral density. Furthermore, a fluoride-containing medication is currently being used for the treatment of osteoporosis.

Magnesium vs. Osteoporosis.

Magnesium has an important role in bone metabolism, as a significant proportion of the body's magnesium (60%) is found in bones. In addition, magnesium regulates active calcium transport. Research has been conducted on the effect of magnesium upon bone mineral content; however, they have not examined magnesium supplementation alone (the supplements contained other minerals such as calcium, phosphorus and boron). One study has shown that dietary magnesium is correlated with bone mineral density in postmenopausal women. Refer to Table XXIV for a description of magnesium-rich foods.

Phosphorus vs. Osteoporosis.

Mature bone mineral actually consists of calcium, phosphorus, oxygen, and hydrogen Collectively, this crystal is referred to as hydroxyapatite. Hence, phosphorus has an integral and important role in bone structure. Phosphorus, however, is another example of an "ironic" mineral. Inadequate intake can result in bone mineral loss (because of a lack of mineral mass), and excessive intake can also result in bone mineral loss (because of excessive urinary calcium excretion). The typical American diet contains more than adequate quantities of phosphorus, and in fact, may be excessive in phosphorus, particularly if high in baked goods, cheeses, processed meats, and soft drinks. These foods contain food phosphorus-containing food additives which can account for 20% to 30% of dietary phosphorus. Review Table XXV for a listing of phosphorus-rich foods.

Table XXV. Phosphorus-Rich Foods.

Phosphorus-Rich Food	Phosphorus Content (mg/serving)	Serving Size	Calories per Serving
Cereals, 100% bran	801	1 cup	178
Yogurt, plain, low-fat	326	1 cup	144
Milk, skim	247	1 cup	86
Cheese, Swiss, pasteurized, processed	216	1 oz	95
Waffles, frozen	135	1 each	218
Cheese, Mozzarella, part skim	131	1 oz	72
Cola	44	12 oz	152

Copper, Manganese and Zinc vs. Osteoporosis.

Several trace elements, such as copper, manganese and zinc are necessary cofactors for enzymes which are essential in bone metabolism. One study has shown that supplementation of these minerals, along with calcium, significantly increases spinal bone mineral density in postmenopausal women.

In summary, because of the integrated role of many nutrients in osteoporosis prevention, many calcium supplements contain a blend of many of the minerals listed above and certain vitamins.

Foreign Body Killer?

Zinc

The immune system is important for eliminating foreign bodies and potentially disease-causing microbes. In addition, new theories are emerging which suggest that certain types of cancers have an origin in viruses. Hence, the immune system becomes important in preventing acute infection and chronic disease. What is the role of zinc? Research has clearly demonstrated that a deficiency of this mineral results in lowering of the number of immunity cells. However, consuming zinc in excess of RDAs does not provide additional benefit to immune system. Refer to Table XIV on page 36 for a listing of zinc-rich foods.

CHAPTER SEVEN

Where Are They?

Food Minerals.

As you have probably deduced from earlier sections of this book, food sources are the preferred resource for nutrients, including minerals. But why?

The Cephalic Phase.

Secretions in the digestive tract, including stomach acid and products from the pancreas help separate minerals from chelators, keep them in their elemental form, and enhance absorption. These secretions are stimulated when we eat real food (remember the "cephalic phase?"). Hence, you can enhance their bioavailability by getting minerals from food sources or at least by taking your supplements with a meal. Let your body work for you.

Nutrient Density.

If a food is a rich nutrient source for one or a group of nutrients, then most likely it is also a good source for other essential nutrients. For instance, whole grains are an excellent source of zinc, magnesium, and iron. They are also rich in many of the water soluble vitamins and dietary fiber and other phytochemicals*.

* The term "phyto-" is derived from "plant." Phytochemicals are compounds found in plant foods (i.e., grains, cereals, fruits, and vegetables). These chemicals have spurred much interest because of their ability to prevent and treat disease. This is why vegans and populations that eat more plant foods consistently live longer and have a lower incidence of many chronic diseases, such as coronary heart disease, hypertension, cancer, diabetes, and obesity.

For more information about mineral-rich food sources, please refer to the section titled, "Minerals As Nutraceuticals." That section reviews specific minerals and their influence upon optimal health and disease prevention. Food sources for specific minerals are presented in this section.

Food Storage.

Some storage tips to maximize the mineral (and vitamin) content of foods:

- ❑ For fresh food, purchase only that amount that can be eaten within a few days.
- ❑ Store refrigerated and frozen foods under appropriate conditions (40°F and 0°F, respectively). Store canned and dry goods in a cool, dry place.
- ❑ Store canned or frozen foods for a maximal time of 3 to 5 months.
- ❑ Store bulk dried goods (i.e., beans, noodles, rice, flour) in dark containers or in the refrigerator.

Food Preparation.

Some preparation tips to maximize the mineral (and vitamin) content of foods:

- ❑ Thaw frozen meat, poultry, and fish in the refrigerator before cooking.
- ❑ Cook frozen vegetables without thawing.
- ❑ Remove only the outer layer of skin on produce.
- ❑ When cooking food, minimize: (1) cooking water, (2) time and temperature*, (3) chopping or dicing.
- ❑ Cook vegetables just until crisp to tender.
- ❑ Use celery leaves, parsley, and other leftover vegetables for soup stock.

Use leftover liquids for sauces, soups, stews, or cooking water for cereals, rice, or noodles.**

* <u>Always cook for sufficient time and temperature</u> to insure that microbes which cause food-borne illnesses are completely destroyed and food is safe. This is particularly critical for the very young and the elderly. Many foods which are susceptible

to harboring these microbes provide time and temperature guidelines on their package.

** Some leftover liquids may be high in fat, particularly if used as a basting stew for meat. Simply refrigerate the liquid until a solid layer of fat forms at the surface. Remove this fat layer and use the bottom portion as recommended above.

CHAPTER EIGHT

All About Colloidal Minerals

The popularity, interest and wonderment about colloidal minerals warrants a special section. This section provides information on the nutritional content of various colloidal mineral supplements that were available at the time of this writing. Note that the nutrient information presented is derived from manufacturer's labels and has not been validated by an independent laboratory.

- *Product A contents:*
  Colloidal mineral mixture
  Deionized water, calcium, chromium, copper, iodine, lithium, magnesium, manganese, molybdenum, potassium, selenium, silicon, silver, vanadium, zinc.

- *Product B contents:*
  Colloidal mineral mixture
  Calcium, chromium, copper, iodine, lithium, magnesium, manganese, molybdenum, potassium, selenium, silicon, silver, vanadium, zinc in deionized water.

  Critique of Products A and B.
  Ironically, the deionized water used in these mineral suspensions has been filtered to remove minerals!
  The actual amounts of the minerals listed in these Product A and B are not provided on the label; the minerals appear to be listed in alphabetical order, rather than in order of descending quantity, as is the Food and Drug Administration (FDA) standard for food labels.

- *Product C contents:*
  Silicon, boron mineral mixture
  Boron at 2 ppm and silicon at 1000 ppm in deionized water.

  Critique of Product C.
  It is still questionable whether boron is an essential nutrient, although it may be useful for the prevention of osteoporosis. It has been rumored that silicon and boron aid in the absorption of other minerals, including calcium, although this idea remains unsupported by the research.

- *Product D contents:*
  Enhanced calcium liquid supplement
  One daily suggested serving (1 tsp) contains:
  Calcium citrate and calcium carbonate: 500 mg
  Magnesium carbonate and magnesium citrate: 250 mg
  Silicon (colloidal silica): 6 mg
  Boron (potassium borate): 1.0 mg

  Critique of Product D.
  This product appears to be designed for the prevention of osteoporosis. One daily suggested serving, however, does not provide the USRDA for calcium, and at this dosage, calcium-rich foods still remain an important part of the diet.

- *Product E contents:*
  Chromium, vanadium mineral mixture
  Chromium: 0.2 ppm
  Vanadium: 0.2 ppm

  Critique of Product E.
  It has been rumored that chromium and vanadium aid in the metabolism of glucose, and this supplement appears to be designed for this purpose. Chromium, when injected as part of the glucose tolerance factor does appear to reduce hyperglycemia in diabetic rats, however the chromium in this product is most likely not in this form. Furthermore, there is inconclusive evidence that vanadium plays a role in glucose metabolism or in alleviating diabetic hyperglycemia.

- *Product F contents:*
  Silver colloidal
  Silver: 5.0 ppm in deionized water

  Critique of Product F.
  There are many health claims associated with colloidal silver supplements, however, all of these remain unproved. Silver is a potentially toxic mineral, and there have been reported cases of toxicity from excess environmental and dietary exposure.

CHAPTER NINE

Parting Comments

As we emerge from our journey of enlightenment, we realize that minerals are diverse in their biological roles and particularly important as directors in the dance of metabolism. We have developed a new respect and recognition for minerals, particularly with regard to their potential as nutraceutical soldiers in the war against disease.

Because of their important roles in metabolism and well-being, optimal nutriture and bioavailability of these essential nutrients has become an ever increasing concern, if not at an alarmist rate. Resultantly, many health claims have been exaggerated and unfounded about the bioavailability and physical-chemical form of mineral supplements.

As scientific discovery advances the field of mineral nutrition, one should expect to find ever more diverse roles of these essential nutrients, particularly in the realm of disease prevention and treatment.

BIBLIOGRAPHY

Antonios TF, MacGregor GA. Salt intake: potential deleterious effects excluding blood pressure. J. Hum. Hypertension. 1995. 9(6):511-5.

Barrett-Connor E, Chang JC, Eldelstein SL. Coffee-associated osteoporosis offset by daily milk consumption. The Rancho Bernardo Study. J. Am. Med. Assoc. 1994. 271(4):280-3.

Clancy SP, Clarkson PM, DeCheke ME, Nosaka K, Freedson PS, Cunningham JJ, Valentine B. Effects of chromium picolinate supplementation on body composition, strength, and urinary chromium loss in football players. Int. J. Sports Nutr. 1994. 4(2): 142-53.

Clark K, Sowers MR. Alcohol dependence, smoking status, reproductive characteristics, and bone mineral density in premenopausal women. Res. Nurs. Health. 1996. 19(5):399-408.

Couzy F, Keen C, Gershwin ME, Mareschi JP. Nutritional Implications of the interactions between minerals. Prog. Food Nutr. Sci. 1993. 17:65-87.

Davies S,McLaren HJ, Hunnnisett A, Howard M. Age-related decreases in chromium levels in 51,665 hair, sweat, and serum samples from 40,872 patients—implications for the prevention of cardiovascular disease and type II diabetes mellitus. Metab. Clin. Exp. 1997. 46(5): 469-73.

Dreosti IE. Recommended dietary intakes of iron, zinc, and other inorganic nutrients and their chemical form and bioavailability. Nutrition. 1993. 9(6):542-5.

Eaton-Evans J. Osteoporosis and the role of diet. Br. J. Biomed. Sci. 1994. 51(4):358-70.

El-Bayoumy K, Upadhyaya P, Chae YH, Sohn OS, Rao CV, Fiala E, Reddy BS. Chemoprevention of cancer by organoselenium compounds. J. Cell. Biochem. Supp. 1995. 22: 92-100.

Evans GW, Pouchnik DJ. Composition and biological activity of chromium-pyridine carboxylate complexes. J. Inorg. Bioch. 1993. 49(3): 177-87.

Felson DT, Zhang Y, Hannan MT, Kannel WB, Kiel DP. Alcohol intake and bone mineral density in elderly men and women. The Framingham Study. Am. J. Epi. 1995. 142(5):485-92.

Flicker L, Hopper JL, Rodgers L, Kaymakci B, Green RM, Wark JD. Bone density determinants in elderly women: a twin study. J. Bone Min. Res. 1995. 10(11):1607-13.

Frolich W. Bioavailability of micronutrients in a fibre-rich diet, especially related to minerals. Eur. J. Clin. Nutr. 1995. 49(3S):S116-22.

Fung MC, Bowen DL. Silver products for medical indications: risk-benefit assessment. J. Toxciol. 1996. 34(1):119-26.

Hallmark MA, Reynolds TH, DeSouza CA, Dotson CO, Anderson RA, Rogers MA. Effects of chromium and resistive training on muscle strength and body composition. Med. Sci. Sports Exer. 1996. 28(1): 139-44.

Harris SS, Dawson-Hughes B. Caffeine and bone loss in health postmenopausal women. Am. J. Clin. Nutr. 1994. 60(4):573-8.

Holbrook RL, Barrett-Connor E. A prospective study of alcohol consumption and bone mineral density. Clin. Res. Ed. 1993. 306(6891):1506-9.

Hulten L, Gramatkovski E, Gleerup A, Hallberg L. Iron absorption from the whole diet. Relation to meal composition, iron requirements and iron stores. Eur. J. Clin. Nutr. 1995. 49(11):794-808.

Karakucuk S, Ertugrul Mirza G, Faruk Ekinciler O, Saraymen R, Karakucuk I, Ustdal M. Selenium concentrations in serum, lens and aqueouis humour of patuients with senile cataract. Acta Ophthal. Scand. 1995. 73(4):329-32.

Kuczmarski RJ. Increasing prevalence of overweight among US adults. J. Am. Med. Assoc. 1994. 207:205.

Laitinen K Karkkainen M, Lalla M, Lamberg-Allardt C, Tunninen R, Tahtela R, Valimaki M. Is alcohol an osteoporosis-inducing agent for young and middle-aged women? Metab: Clin. Exp. 1993. 42(7):875-81.

Laitenen K, Valimaki M. Bone and the 'comforts of life'. Ann. Med. 1993. 25(4):413-25.

Larsen T. Dephytinization of a rat diet. Consequences for mineral and trace element absorption. 1993. Biol. Trace Element Res. 1993. 39(1):55-71.

Lu J, Pei H, Ip C, Lisk DJ,Ganther H, Thomson HJ. Effect on an aqueous extract of selenium-enriched garlic on in vitro markers and in vivo efficacy in cancer prevention. Carcinog.1996. 17(9):1903-7.

Lukaski HC, Bolonchuk WW, Sliders WA, Milne DB. Chromium supplementation and resistance training: effects on body composition, strength, and trace element status of men. Am. J. Clin. Nutr. 1996. 63(6):954-65.

Luoma PV, Nayha S, Sikkila K, Hassi J. High serum alpha-tocopherol, albumin, selenium and cholesterol, and low mortality from coronary heart disease in northern Finland. J. Int. Med. 1995. 237(1):49-54.

MacGregor GA, Cappuccio FP. The kidney and essential hypertension: a link to osteoporosis? J. Hypertension. 1993. 11(8):81-5.

McCarron DA. Role of adequate dietary calcium intake in the prevention and management of salt-sensitive hypertension. Am. J. Clin. Nutr. 1997. 65(2S): 712S-16S.

Mirsky N. Glucose tolerance factor reduces blood glucose and free fatty acids legels in diabetic rats. J. Inorg. Bioch. 1993. 49(2): 123-8.
Murray MT Encyclopedia of Nutritional Supplements. Prima Publishing. 1996.

Narhinen M, Cernerud L. Salt and public health—policies for

dietary salt in the Nordic countries. Scand. J. Primary Health Care. 1995. 13(4):300-6.

Naghii MR, Samman S. The role of boron in nutrition and metabolism. Prog.Food Nutr. Sci. 1993. 17(4):331-49.

Nordin BE, Need AG, Morris HA, Horowitz M. The nature and significance of the relationship between urinary sodium and urinary calcium in women. J. Nutr. 1993. 123(9):1615-22.

Odvina CV, Safi I. Wojtowicz CH, Barengolts EI, Lathon P, Skapars A, Desai PN, Kukreja SC. Effect of heavy alcohol intake in the absence of liver disease on bone mass in black and white men. J. Clin. Endocr. Metab. 1995. 80(8):2499-503.

Owen RW, Weisgerber UM, Carr J, Harrison MH. Analysis of calcium-lipid complexes in faeces. Eur. J. Canc. Prev. 1995. 4(3):247-55.

Patterson BH, Levander OA. Naturally occurring selenium compounds in cancer chemoprevention trials: a workshop summary. Canc. Epi, Biomark. Prev. 1997. 6(1): 63-9.

Pence BC. Role of calcium in colon cancer prevention: experimental and clinical studies. Mut. Res. 1993. 290(1):87-95.

Pietinenn P, Vartianinen E, Seppanen R, Aro A, Puska P. Changes in diet in Finland from 1972 to 1992; impact on coronary heart disease risk. Prev. Med. 1996. 25(3):243-50.

Riggs BL, Melton LJ. The prevention and treatment of osteoporosis. New Eng. J. Med. 1992. 327:620.

Saltman PD, Strause LG. The role of trace minerals in osteoporosis. J. Am. Coll. Nutr. 1993. 12(4):384-9.

Schauss A.G. Minerals, Trace Elements and Human Health. Tacoma WA: Life Sciences Press. 1996.

Schauss AG. An analysis of colloidal mineral claims. Parts 1 and 2. Health Counselor. 1997. IMPAKT Communications, Inc.

Schwarz B, Bischof HP, Kunze M. Coffee, tea, and lifestyle.

Prev. Med. 1994. 23(3):377-84.

Tranquilli AL, Lucino E, Garzetti GG, Romanini C. Calcium, phosphorus and magnesium intakes correlate with bone mineral content in postmenopausal women. Gynec. Endocrin. 1994. 8(!):55-8.

Trent LK, Thieding-Cancel D. Effects of chromium picolinate on body composition. J. Sports Med. Phys. Fitness. 1995. 35(4): 273-80.

Volpe SL, Taper LJ, Meacham S. The relationship between boron and magnesium status and bone mineral density in the human: a review. Magnes. Res. 1993. 6(3):291-6.

Wardlaw GM. Putting osteoporosis in perspective. J. Am. Diet. Assoc. 1993. 93(9):1000-6.

Whiting SJ. The inhibitory effect of dietary calcium on iron bioavailability: a cause for concern? Nutr. Rev. 1995. 53(3):77-80.

Wienk KJ, Marx JJ, Lemmens AG, Brink EJ, Van Der Meer R, Beynen AC. Mechanism underlying the inhibitory effect of high calcium carbonate intake on iron bioavailability from ferrous sulphate in anaemic rats. Br. J. Nutr. 1996. 75(1):109-20.

Wilson BE, Gondy A. Effects of chromium supplementation on fasting insulin levels and lipid parameters in healthy, non-obese young subjects. Diab. Res. Clin. Prac. 1995. 28(3): 179-84.

Zhou JR, Erdman JW Jr. Phytic acid in health and disease. Crit. Rev. Food Sci. Nutr. 1995. 35(6): 495-508.

GLOSSARY

Adequate:
The level of nutrient intake estimated to maintain the health of the general population.

Antinutrients:
Dietary components which impair the absorption and metabolism of other nutrients.

Cephalic phase:
The physiological response which is stimulated by the anticipation or act of eating.

Chelator:
A chemical compound attached to a mineral; these are usually "organic," or carbon-containing.

Colloidal minerals:
A mineral mixture suspended in liquid, usually water.

Dietary fiber:
A component found in plant foods (i.e., cereals, grains, fruits, vegetables, legumes, nuts, seeds), which is indigestible by the human intestine.

Electrolyte:
Minerals which do not have a positive or negative charge, but have the potential to become positively or negatively charged.

Enzymes:
Proteins that aid in metabolism; proteins which can control and enable chemical reactions. Some enzymes require specific minerals for their function.

Essential nutrient:
A chemical compound found in foods, which either cannot be made by the body, or cannot be made in an adequate amount to meet the body's needs.

Estimated Minimum Requirement (EMR):
The estimated nutrient intake which is thought to prevent

nutritional deficiency.

Estimated Safe And Adequate Daily Dietary Intake (ESADDI):
The estimated nutrient intake which is thought to provide adequate nutrition. Because insufficient research information is available to provide an RDA for these nutrients, the ESSADI provides a range rather than a specified value.

Ion:
Minerals which are positively or negatively charged.

Major mineral:
A mineral with a dietary recommendation at or in excess of 100 mg per day.

Metabolism:
The sum of chemical reactions which occur in the body.

Microgram (ug or mcg):
An amount equivalent to 0.000001 gram or 0.00000035 ounces.

Milligram (mg):
An amount equivalent to 0.001 gram or 0.000035 ounces.

Minerals:
Inorganic nutrients (free of the element, carbon) that allow for enzyme activity and metabolism.

Nutriture:
A term describing how well the body is nourished.

Nutraceutical:
A nutrient with the potential to treat or prevent a disease or condition.

Optimal:
The level of nutrient intake estimated to provide maximum health and performance.

Parts per million (ppm):
An amount approximately equivalent to 0.001 gram or 0.000035 ounces.

Parts per billion (ppb):
An amount equivalent to 0.000001 gram or 0.00000035 ounces.

Primary deficiency:
A nutritional deficiency due to inadequate intake of a specified essential nutrient.

Recommended Dietary Allowance (RDA):
Recommended intakes of nutrients that meet the needs of almost all healthy people of similar age and gender.

Recommended Dietary Intake (RDI):
The suggested average daily dietary intake for an essential nutrient.

Secondary deficiency:
A nutritional deficiency due to a condition (i.e., disease, impaired bioavailability) other than inadequate intake of the specified nutrient.

Trace mineral:
A mineral with a dietary recommendation less than 100 mg per day.

USRDA:
The RDA for adults, male or female, whichever is greater.

INDEX

ABOUT BIOMED

Biomed is an organization that provides health care professionals with the latest scientific and clinical information, enabling these professionals to provide better care for their patients. The information is provided by means of live seminars and home-study courses.

Biomed operates nation-wide in the United States. The company also operates internationally. Biomed is a subsidiary of the Institute for Natural Resources (INR).

For more information about Biomed-INR seminars and home-study programs, please contact Biomed-INR. The address and telephone numbers are listed below.

PUBLISHER'S NOTE

Although every effort has been made to ensure the accuracy and completeness of the information, the publisher cannot be responsible for any errors or omissions.

Biomed
5801 Christie Avenue, Suite 400
Emeryville, California 94608 USA
Tel: (510) 450-1650
Fax: (510) 652-1859

HOME-STUDY

BIOMED HOME-STUDY

COURSE #5100

COURSE TITLE:
"Minerals and Health"

TEXTBOOK:
"Precious Metals"

AUTHOR:
Carol Jong, Ph.D., R.D.

CONTINUING EDUCATION CREDIT: Two hours
PROCESSING FEE: $2.00

To receive credit for this course, you must:

1) complete the Examination on page 3 and achieve a passing score of 70%.

2) clearly fill out the Questionnaire and Registration Form on page 4.

3) return pages 3 and 4 -- along with the $2.00 processing fee (payable by check or Visa/MasterCard) -- by December 31, 1998. Please send all materials to:

Biomed
P.O. Box 4218
Berkeley, California 94704-0218 USA
(510) 450-1650

Note: Please make check (if applicable) payable to: Biomed. A certificate verifying course participation will be sent. Please allow four to six weeks for processing.

INSTRUCTIONAL OBJECTIVES:

Upon completion of this course, the health professional will be able to:

1) understand the role of minerals in metabolism and the body's functions.

2) analyze the current knowledge in the area of minerals and their roles as nutraceuticals in the treatment and prevention of diseases.

3) critically analyze current mineral supplements as an informed consumer.

4) understand the relationship among minerals, nutrients and medications and their interactions in metabolism.

This course is sponsored by Biomed, the home-study division of the Institute for Natural Resources (INR).

Biomed is a scientific organization dedicated to research and education in the fields of health and medicine. Biomed has no ties to any commercial organizations, does not solicit or receive any grants or gifts from any source, and has no connections with any religious or political entity. Its mailing address is:

Biomed
P.O. Box 4218
Berkeley, California 94704-0218 USA

The continuing education credits for any Biomed course are awarded by Biomed's parent organization, the Institute for Natural Resources (INR). All course approvals are held directly by the Institute.

Approvals

The table listed below lists, for various professional groups, the organizations that have accredited the Institute for Natural Resources (the parent company for Biomed) as a sponsor of continuing education in nursing by the American Nurses Credentialing Center's Commission on Accreditation. The Institute for Natural Resources (INR) is approved by the American Psychological Association to offer continuing education for psychologists. INR maintains responsibility for the program. You may wish to check with your own licensing board to determine if the accreditations listed below are acceptable to your board.

Professional Group	Accrediting Organizations
Registered Nurses, Licensed Practical Nurses, & Licensed Vocational Nurses	American Nurses Association/American Nurses Credentialing Center California Board of Registered Nursing (Provider #06136) Florida Board of Nursing (Provider #27I1118) Iowa Board of Nursing (Provider #288)
Dentists, Dental Hygienists, Registered Dental Assistants	Academy of General Dentistry (national office) California Board of Dental Examiners Florida Board of Dentistry Indiana Board of Dentistry
Psychologists	American Psychological Association
Dietitians	American Dietetic Association (Application made January 1998)
Pharmacists (Initial release date: January, 1998)	INR is approved by the American Council on Pharmaceutical Education (ACPE) as a provider of continuing pharmaceutical education. (Universal program number is 751-000-98-007-H01)
Social Workers	American Psychological Association California Board of Behavioral Science Florida Agency for Health Care Administration Illinois State Dept. of Professional Regulation

Examination

True-False:

1. With the exception of iron, minerals are best absorbed in the presence of food. T F
2. Major minerals are physiologically more important than trace minerals. T F
3. An excess of one mineral can decrease the bioavailability of another mineral. T F
4. Current research has shown that individuals consuming a moderate amount of alcohol (1 oz. per day) have greater bone densities than individuals consuming less alcohol. T F
5. Chelated minerals are better absorbed than non-chelated (elemental) minerals. T F
6. Usually the quantity necessary to achieve optimal nutriture is greater than the quantity necessary for adequate nutriture. T F
7. Calcium may be an effective nutraceutical for the prevention of colon cancer. T F
8. The "cephalic phase" describes the physiological response elicited by a toxic amount of aluminum in the diet. T F
9. The statement, "U.S. Pharmacopia" indicates that a nutritional supplement satisfies purity standards. T F
10. Chromium picolinate is very bioavailable. T F
11. The chelator, picolinate, has been shown to cause abnormalities sin cell cultures. T F
12. "ND" means that the level of a nutrient is too low as to be detectable. T F
13. Selenium can spare the need for vitamin D. T F
14. Selenomethionine is the biologically active form of selenium in humans. T F
15. Zinc lozenges have been shown to be effective at reducing sore throat symptoms. T F

Multiple choice:

1. Minerals aid in metabolism by:
   a) converting electrolytes into ions
   b) increasing the bioavailability of amino acids
   c) altering the fluid balance in the body
   d) activating enzymes
   e) all of the above

2. Through perspiration, the weekend athlete loses significant quantity(s) of:
   a) sodium
   b) potassium
   c) chloride
   d) fluid
   e) all of the above

3. Calcium may be an effective treatment for hypertension because this mineral:
   a) causes arterial dilation
   b) increase the urinary excretion of sodium
   c) enhances weight loss
   d) none of the above
   e) all of the above

4. Colloidal minerals may contain______ which can be toxic to the body.
   a) aluminum
   b) arsenic
   c) lead
   d) sodium
   e) all of the above

5. Using the "Rule of 300," the calcium intake of an individual consuming a balanced diet and one cup of yogurt each day is:
   a) 300 mg
   b) 600 mg
   c) 900 mg
   d) 1200 mg
   e) none of the above

Biomed
Home-Study Course
#5100 (2 Hours)

Questionnaire

a)	The scientific information was clearly presented.	1 2 3 4 5
b)	I acquired useful, new knowledge from this course.	1 2 3 4 5
c)	I found this course interesting.	1 2 3 4 5
d)	This course met my professional objectives.	1 2 3 4 5
e)	The author's writing style was effective.	1 2 3 4 5
f)	I would take another course by this author.	1 2 3 4 5
g)	I would recommend this course to a professional colleague.	1 2 3 4 5
h)	How many hours did it take you to complete this course? ____________hours.	
i)	Do you have any suggestions for future home study courses?	

____

____

j) Other comments:

____

____

Signature ______________________________ Date ______________

Registration Form

Name ____

Home Address ____

City __________________________ State ______ Zip Code ____________

Day Phone (   ) ____________________ Evening Phone (   ) ____________________

Profession ____________________ Prof. License # ____________________

Please enclose the processing fee with this form. Check method of payment.

- ❑ Check for the processing fee of $2.00 (Make payable to Biomed)
- ❑ Charge the amount of $2.00 to my ❑ Visa ❑ MasterCard

Card Number:______________________________Exp. Date:__________

Signature ____

Home-Study Courses:

Biomed has prepared a wide range of home-study courses, books, tapes, and other materials for health professionals. To obtain more information on these courses and products, please contact:

Biomed Publishing Department
P.O. Box 4218
Berkeley, CA 94704-0218
Phone: (510) 450-1650
Fax: (510) 652-1859

☐ **Yes! Please send me more information about Biomed materials**

(Please Print)

Name:____

Home Address:_________________________________

City:________________________________

State:______________Zip Code:__________

Day Phone:___________________________

Evening Phone:________________________

Fax:________________________________